The Anatomy of Love

The Five Elements of Love

HUGO BRADFORD

Acknowledgments

As I sit here, humbled and reminiscent of this book's journey, I have before me pages of a manuscript with red markings and meritorious notes from my editor, The Persnickety Proofer that I can call estimable work.

This journey began as a longing for drowned emotional expression. It patiently started taking form with every word, sentence, and paragraph written a day at a time, culminating in what you hold in your hands, my debut, *The Anatomy of Love*. Yet, none would have been possible without the love and support of many people. First and foremost, I'd like to thank and praise my beloved parents, Joe and Aracely Bradford, and my sister, Mary Beyonce, whose love and words of affection I have carried with me through the passage of the years since my more solemn teenage years. Standing on a pedestal above all, I want to thank my revered mother, a woman whose grit against life's many obstacles never wavers. Your unconditional love and wisdom has touched me since my first memories—where would I be without you? The development of my mind and intense personality were assigned to the only person capable of prevailing in such a job—my mother, Aracely Bradford.

But what these pages hold, and what's to become the next chapter in my life wouldn't have been possible without my closest friends, who've been with me through thick and thin. They were there to praise my accolades, challenge me, make me aware of my flaws, and give me strength when I had none. Sebastian Castaneda, you're more than my best friend; you're my brother, one I am terribly grateful to have and privileged to call business partner. I'm eternally thankful for your support, insight, and friendship, as few minds can understand me at such depth and ponder with me about the many marvelous wonders of the universe. Our friendship often makes me wonder about the fortuitous nature of our paths. One born on a Sunday

(day of the sun) and the other on Saturday (day of Saturn), corresponding to our signs. Perhaps sheer coincidence, or maybe fate, meant for the Sun and Saturn to rule the world.

Sketch Baptisme is another brother whose friendship supersedes its pronunciation. You are a talented individual who holds a standard of life that quells the adversities of your past, at which I can only stand in awe. I am honored to call you my friend, and much obliged of your beholden advice, competitiveness, and unquestionable support.

Maritza Aguilar, my Mari, a lioness, and a lion, you and I, whose ardent natures sometimes claw at each other's pride when in the same habitat for extended times. But most of all—you are an angel. Such that my gratitude cannot justly be put into words, as I'm endlessly indebted to you for your kindness, affection, support, and tolerance for me. You have challenged me and been there for me in moments when only a tiny few would. From our vulnerable moments to our laughs and concerts in the car, my love and respect for you are absolute—thank you.

But this modest acknowledgment wouldn't be complete without commending Sarah Weiser, a cherished friend with an immense heart that radiates warmth hundreds of miles. I'm thankful for your affection, wisdom, and head-strong resolve. These traits can only be complemented by your significant other, Jeremy, for whose relationship I hold only admiration and utter respect. I also want to thank Cinthia Franco, Armon Tillman, Whitney Wilkinson, Tanya Ortega, Elliott Gular, my lifelong friends Johnross Parra, Ania Ptak, and Jaime Lara, along with Dana Palmeri, my cousin Mireya Holguin, and my spoiled and beloved niece Ingrid Bustamante. My niece, who I hope to inspire to see that anything's possible when you set your mind to it. Karla, Armando, Ian, and of course, my treasured and beautiful Caesar and Bliss Beyonce

Thank you from the most profound part of my heart, for it is because of you I am a wiser and better man today. These pages hold a piece of each of you to be shared with the world, breaking the laws of physics traveling through the dimensions of time. Because of you, I have planted what I hope will be my first seed of many in the world of literature that I love and respect.

Contents

Introduction

I couldn't be happier you're here reading this page because I believe it to be necessary before the voyage ahead. Furthermore, I'm glad you're here because it means it was all worth it—the pain, the late nights, the long workdays. The very fact that you have this book in your hands makes my position as a science writer official—a lifelong dream of mine. But most of all, it brought *you* here—the only purpose of this book's existence.

Perhaps getting here wasn't easy for you either, as you might have a telling journey of your own. But it seems that's the norm in the world of love—beauty, pain, drunken happiness. But encapsulating it all is an overwhelming emotion. It is this polarity in which love operates that drew my inquisitive nature to pose the question myself —what is love?

In an array of means, life operates in a constant paradoxical polarity: good and evil, life and death, night and day, dominance and weakness—survival of the fittest, black holes that swallow everything around them, even light, yet life couldn't exist without them. The trend continues. Males and females, love and…? What is the polar opposite of love? If you're like most, you might have answered hate, yet I disagree. Hence the partial catalyst to writing this book, as I believe the polar opposite of love to be *fear*.

Amid life's riddles, I noticed a clash of titanic forces and forget about Goliath. It's an unstoppable force meets an unmovable object—women vs. men. A detail I not only wish to stop but also find nonsensical. Throughout my life, I've always been given three primary compliments: "you are handsome," "you are intelligent," and "you're a good guy." Thus, as a kid, beautiful women in my life—my grandma, mother, aunts, friends of the family—filled my head with tales and illusions of grandeur, all while I heard the adult males in the family, loquacious actors, and TV hosts quipping how they didn't understand women. To which I thought, *what could be so hard to understand? I'm surrounded by women and love them all.*

As I climbed the ladder of my teenage years and familiarized myself with the alluring traits of the opposite sex, I approached my long-awaited eighteenth birthday with excitement. I couldn't wait to get to college and relive all the stories I watched in movies and read in books— *The Notebook,* Jack and Rose, Chris Ostreicher (a.k.a. Oz), and Heather. It wasn't long before I had my first official girlfriend, just like it wasn't long before she wasn't anymore; cause to someone's shiny, bright-red, new Mustang. Then another girlfriend, to a shiny new sports car. Two concurrent factors proved to be right: one, out of the three recurring compliments (handsome, intelligent, good guy), at least one of them wasn't working. And two, it appeared I didn't understand women after all, either.

Due to all the factors as mentioned above, I decided to change, and even though it wasn't what I envisioned ideally, I got rid of the *good guy* and got myself a Mustang. I then began noticing a second pattern—women liked me. I was popular, and the less I cared, the more it elevated that status. I heard compliments such as "you're sexy," "I can't believe you don't have a girlfriend," and "you should call me" came to me in slipped folded napkins, flirtatious looks, and even by directly approaching me when I was sipping on a deli smoothie at the mall. Perhaps they were all consequence to the sound of a loud double exhaust, or maybe my newfound confidence. Who knows, but the more I was asked about a girlfriend, the less I wanted one. After all, at the time, I had what I wanted—attention and sex without the liabilities of a relationship. At least, that was what I thought at the time, but I was in my early and mid-twenties, a vastly different man than I am now.

As the years passed, I met a girl I liked. She understood me, supported me, and loved me, but also the music, outings, watching movies, and traveling had a different meaning. It felt like I was living a movie, full of potent emotions. But something happened—me. I had grown accustomed to being a less caring guy. I often had an attitude and frequently pretended I didn't care. As I saw it, it was those personality traits that had gotten me the successful romantic streaks and ultimately landed me my girlfriend. *So why change?* I thought. I had officially and effectively graduated into a professional, pretentious, two-faced douchebag who spurred around like his sh** didn't sink. I convinced myself she wasn't the one for me, she didn't have my drive and goals, that in all my "might" I deserved more. The truth was, I didn't *deserve* more, I *wanted* more. So, as the old Chinese proverb says, *"Be careful of what you wish for…"* which I wasn't, and consequently triumphed in my wish of being single again and additionally ruining what might have been the love of my life. Shortly after the painful fog of the breakup cleared and my mind was a bit soberer, I couldn't help my guilty thought about my ex-girlfriend: *if she believed in love, how will she change now?*

Years of introspection, growth, and arduous research followed. I was determined to find the Hugo I had lost, somewhere in the abyss of a whiny, bitter past, floating along with the memories of female faces that whispered *what might have been?* Determined to change, I did, and as you read on, we approach the moment that climaxed to getting us to this very page. I met a woman who enthralled me from the very moment my eyes landed on her. We slowly started moving fast, and we began making memories even quicker, and then—it disappeared. Though I had tried to be the best version of me I had ever been, the best boyfriend as I saw fit, I became intimidated by my personality and complacent aim to please.

This paragraph is the last climatic capstone that got you and me here, to the story you're about to hear. A story built from the many lessons I have gathered from introspection, personal knowledge, talking to pros, interviewing a vast number of women, and research from different fields of science like psychology, anthropological biology, and social science.

As I questioned, pondered, and coped with the pain, I reached a pleasant resolve—finding the middle. I deduced the obvious—we're always reacting to our past and, consequently, ruining our future by not engaging correctly in our present. It's not about women vs. men, or men vs. women, nor is this a grievance about women. On the contrary, every word within these pages is intended for the woman reading this right now. We can see it as a proper and friendly sit-down between you and me. One intended to provide insightful perspective from a man's viewpoint with comprehensive data to help us both.

I want to be who I truly am. I want to be the alpha male I am but leave the pretentious jerk behind. I want to embrace all of my masculinity without fearing a woman's wrath in my moments of sensitivity or vulnerability. I want us to stop paying for the bad experiences, the neglectful voices, the jerks, and the divas, the machos and the feminazis, the mistakes of people from the past, for the people we meet today. I want for us to stop reacting to the pain of yesterday and continuously living in our memories without cherishing the present for the possibilities of tomorrow.

But for that, I need for us to get on the same page. So please do me a favor. Bring down whatever walls or barriers that might exist between you and me, and I promise to do the same. First, let us recognize the difficulty of listening to what might seem like a judgment at first glance from the voice of a man. Do me a *huge* favor and keep this in mind, it's not a man entering or intruding into *your* world, but instead you *entering* a man's mind. Yet, if I told you that all of the content you have in your hands was easy to digest,

I'd be lying to you—and that I will not do. But what you can rest assured is that it is all without judgment.

On the contrary, raised by a single mother, this is my purest sign of respect for the love I have for all women. But I wouldn't be able to accomplish that without speaking the truth. Yet, speaking the truth sometimes requires us to run the risk of a few getting offended, despite that being the last of my intentions. But I can also promise you that whatever point I make will be backed by facts. Also, I hope this book will be a lot more than that, a bit of humor, some tough love, and an overall emotional and intellectual roller-coaster ride.

Thus, let me take you on this journey through time. Let's embark on a voyage through the sciences, stories with many voices, and the minds of men. Therefore in the name of wiser men and women, in the name of this esoteric word *love*, in the voice of reason, let's get started. We have much to learn.

PART ONE

The Science of Love

My first name is William,
then the J follows the name with a dot.
But my last name's not Blatty,
And though I don't pen down the horror,
I am an explorer, and of the known upper story,
I happen to know and study a lot.

ONE

Defining Love

> Love is the one thing we're capable of perceiving that
> transcends dimensions of time and space. Maybe we should trust
> that, even if we can't understand it.
> – Dr. Brand, Interstellar, 2014

"I think we should break up."

Just moments before that, I thought I was happy. It wasn't the first time my girlfriend and I had been through this, but even then, I was surprised, sad—*afraid*. Even then, my painfully naive reply was, "I can't tell you what to do. All I'll say is I love you, I understand you, I respect you and accept you." I kissed her on the forehead and left.

* * *

Three years earlier

Summer 2017

Just when new relationships were blossoming, mine was ending. I failed to keep what may have been the love of my life. The chemistry, the connection, the support, the intimacy, and the love was all there, but what I sought on paper—the success, the goals, the career path—was not. I effectively succeeded in ending my relationship and breaking my own heart. My ambition made me quite a different person than the one I was to later become. And who I became was vastly different from who I am today.

Consequently, I set out on a path of introspection, challenge, and self-growth for two years. I decided not to be in a relationship until I became a version of myself worthy of giving to the right lady. Someone I thought worthy of having the new me—my success, my new body, my mind. I wanted my next relationship to be my last (don't we all?).

When my last girlfriend and I met, on paper she was everything I wanted—attractive, smart, and had a career path worthy of praise. I liked the validation she brought to my life and the things for which I had worked so hard. At first, the laughs, the sex, the emotional intimacy felt like it was all there. So came the trips, the outings, and late nights bathed in wine, along with meeting friends, family and then came—love.

Just as quickly as time moved while we were having fun, only nine months later, it was gone.

I managed to ignore the warning signs of the image I built in my mind. The new me had simply decided to put full effort into my new relationship. Yet, after I decided to be respectful, supportive, loving, and kind, I succeeded in failing one more time. In the end, one can usually see the signs. But what signs? Two months passed by without a single discussion, that weekend, as usual, we hung out with her friends, a week where I borrowed a book from her friend to an evening in which I simply thought I was happy.

Earlier this evening, we walked to get groceries together. When the bags of alcohol I bought for us to enjoy together during the oncoming pandemic were heavier than we expected, I offered to run back to the apartment to get her car. Once we made it back to her apartment, I bought a movie for us to watch. However, with a half-empty glass of wine, a confusing language exchange caused conflict. For which, I decided to take a breather by taking my laptop to my car to avoid a confrontation. Laughingly joking, I said, "I'll be right back. Don't take the Apple TV down just yet." All to come back from my car to hear her tell me, "I think we should break up."

In the following days, we kept in touch. I hoped to get back together since, all in all, I found the whole thing puerile. After all, we watched the movie and drank some wine. A night where she "forgot" her phone case at my place, and after she came by to pick it up the next morning, we had coffee together, and then, only two days later, we went for a bike ride together for the first time.

The next few days were some of the most humiliating, degrading, and hurtful of my life. Merely three days after our bike ride, I would find out in the most detailed and heinous fashion that she was sleeping with someone else. A gift from that person already sat next to my bottle of wine. At the same time, it's hard to forget her smirk accompanying the atmosphere—after I brought over coffee for both of us.

Her reason for us continuing to engage? "I was talking to you to help you."

Gee, thanks for the favor!

The last talk we ever had? An aloof message from her: *I hope that you continue to choose to stay safe, not only for you but also for your mom and family. Please reach out to your friends or family or the crisis line (1-800-273-8255) if you need to. It is no longer beneficial for us to keep contact; I'm blocking your number.* Kill 'em with [passive-aggressive] kindness, they say.

I replied *I agree.* Not that I had the strength at the moment to walk away. I was so depressed, and I wanted to appeal for her warmth, but my mother assured me it was the best response, and I trusted her resolution.

From that moment on, I never looked back. It was then I understood the difference between *a love story and a story about love.*

* * *

I wish I could tell you this story isn't true, but it is, every aspect of it. So why am I sharing it with you? Because I learned from it, I grew from it, and I found happiness again—and so will you. Because after several relationships, years of introspection, an ocean of confusion, a moment of pain and failure became my journey through science to tackle some of those tough questions we secretly ask ourselves; confident psychology, biology, and social science would have some answers. I researched blog archives, letters, reputable articles from expert voices, and the banks of memories and personal experiences. It was a learning and growing experience that continued with every conversation I had, listening to many women recount their perspective, their frustrations, their vulnerabilities.

I was compelled to find those answers and then share them all with you. Trusting that by the end of this book, you'll have ample understanding and knowledge and the tools to endow you with a new perspective to form your own conclusion, devise your own doctrine, and forge your fortune in the world of love.

So, here's to the wondering minds, to the academic and the analytical, to the ones searching for closure, to the ones searching for love, to the ones with broken hearts. Here's to the artists, to the ink on this page, and the subject in the title. I want to give you answers with a bit of science, a bit of humor, a dash of bluntness, and plenty of psychology to explain it all.

So, here's to the crazy ones, the round pegs in the square holes. Here is the Anatomy of Love.

The Anatomy of Love

Love... A simple, four-letter word that produces over fourteen billion search results on search engines across the internet. It is the third-most searched word worldwide. This simple, four-letter word has been the cause of studies, discussions, debates, and controversy through the ages globally between academics in different fields of science like psychology, biology, social sciences, and more. Still, it has been responsible for some of the most significant art masterpieces of all time. Regretfully, even in our targeted ads.

Yet, everyday people like you and me, along with the rest of the animal kingdom, are not impervious to its enthralling charm, from birds that can have long-term mates, elephants who mourn their dead, and apes that seek the warmth of their mothers. We are, in one way or another, exposed to it every day of our lives. We live love through our romantic relationships, our families, friends, and even pets. Love is everywhere you turn—in your strolls through the park, in your hugs, when you kiss, in every corner of your life, often residing in the deep corners of your memories and thoughts. It is the number one ingredient in the movies we watch (and sometimes cry at), in the books that captivate us, and the songs we sing.

However, despite all of that, today, the word "love" has become complicated to express openly. Humans have—for all our intelligence—an astonishing failure rate in our romantic relationships. Love is a subject so vulnerably intimate that not many are comfortable openly talking about it. We avoid it when we can and act awkward when we can't. But why? Why is it that we fail so much at relationships and love? Why does it hurt so much when love ends, and why do we dilute our pain when it does? Is love something as simple as an evolutionary trait of consciousness and chemical potion in our brains? Or, as Dr. Brand said, does it mean something else, something we can't yet explain?

The Elephant in The Room

One of the great challenges of this world is knowing enough about a subject to know you're right. But not enough about a subject to know you are wrong!
– Neil deGrasse Tyson

Love is a complicated subject, to say the least. The very pronunciation of the word sometimes creates awkwardness. Perhaps, it feels corny or has a dramatic tone, or its sentimental weight is too heavy. Don't think so? Try it out. Go up to a friend, look at them straight in the eye, and say, "I love you."

If you try it out, you may get a kick out of it. You may even receive a reaction while you're at it. In most cases, what you'll find is a hesitant response filled with *hmm, huh?* And *eh*. It goes to show how intimate and vulnerable we can be when merely approached with the subject.

Love and relationships are the subjects that encapsulate most of our lives, yet for some reason (most of us know the reason), we rarely mention them. On the contrary, we avoid them. Evasion, after all, could be where the answer to this riddle lies.

So why do we act the way we do when approached with the subject of love? Pressure applied from programmed social bias? Very good, "You go, Glen Coco!" But in either case, what does it tell us?

"Can't you just tell me already?"

Both guesses highlight fear as the root cause of our avoidance of the topic of love and relationships. The exciting fact in all this is whether the problem is social, psychological, or both, it goes against the very nature of our biology. Let's break the reasoning down into two possible explanations.

Social Bias

Do you remember being in class, and after the professor gave their lecture thinking, *hmm, I didn't understand that very well at all* and then waiting for someone else to raise their hand to express their confusion? And then after a couple of suspenseful moments with the air vent being the only noise in the room and seeing no one raise their hand, following up with the thought, *am I the stupidest person in the class?*

Well, not necessarily. One of the main fears human beings have is stage fright, mainly because we're terrified of sounding dumb and not getting the cumulative consent of people. This phenomenon is otherwise known as the *illusion of transparency,* as Kenneth Savitsky, Ph.D. in Social Psychology from Cornell University explains. Another term that further contributes to providing clarity on the subject is *pluralistic ignorance.*

The **illusion of transparency** is the belief that our internal states are more apparent to others than is the case. Consequently, this is what causes our fear of expressing our thoughts or feelings. Hence, why you hesitate to sing in your car at full blast, fearing you might be the center of attention of the entire highway.

Pluralistic ignorance is when someone acts differently in public as opposed to their personal beliefs because they mistakenly believe that most of their peers might have an opposite opinion. Which also contributes to our fear repressing our emotional expression or our expression about love. See the example below to visualize it better.

| Pluralistic Ignorance & Illusion of Transparency | = Fear ▶ | Conditioning Against Fear ▶ | Repressed Expression |

Suppression, repression, conscious or subconscious, tomato, tomahto—when all is said and done, when we fail to engage on the subject and openly admit it, we succeed at not improving on it. The irony is that love is something we all experience, feel, think about, secretly question, and yearn. If this idea is accurate, then maybe, just maybe, that's the reason why you love to watch some good ol' LMN with tissues on hand and a glass of wine. Or is it that we're permanently destined to tear up at movies alone, get goosebumps when hearing powerful phrases, get shivers from listening to darn good music, and become condemnably accustomed to the occasional traces of moisture on the book page when no one is watching?

Whatever the case might be, it's only you reading this book, with no one to judge you or hurt you. It's just you and me. So, do both of us a great favor for the sake of a better reading experience and remove the barriers mentioned above, as they serve no purpose in our venture. Let the words enter your mind, bare, and unbarred, and continue with the thought that someone, somewhere in the world is reading this page with you, that someone has the same thoughts and questions about this mystery as you do, and in that case, what could be more normal? What could be more necessary?

The Bitter Pill

The world of love can have many phases. Some can be small and go nearly unnoticed—they're the ones that become routine, yet the little things that still matter. You might recognize them as morning pecks from your partner before work, the lazy coffee in bed, the memes shared with meaning only between you two. And then you have the more significant phases—the trips, nights out, the laughter, the decisions together, the gifts, the special hugs. And then, the not so great stages—the disagreements, the arguments, the tears, and the fights.

Finally, there are other phases, the ones that include the moments we'd rather not talk about—the breakups, the sadness, the broken hearts. In all, a roller coaster ride, the ups, the downs, sideways, and beyond.

Regardless of what phase you might be facing, be it that you're searching, be it that you've settled, be it that someone hurt you, one way or another, we are here for the same reason—to find answers, or perhaps out of sheer curiosity.

So now that we're getting better acquainted, and we've already established we watch some of the same movies, read some of the same books, and wonder about the same subjects, the question remains. *Is the love story, the relationship, and chemistry we watch on TV and read about in books really possible?* If you stay optimistic about answering that question, I couldn't be happier for you, because I hope that means you found someone special and that's not very common at all. On the other hand, don't lose hope, as this book is for *you*.

So why is it that you find yourself sometimes asking, *why can't I find someone that gives me more words of affection? Have I settled? Are all men as*****s? Why can't someone stop me at the airport before my flight departs?* Well, TSA, for one, but aside from that, I'm afraid you might find the simplicity of the answer rather disappointing. The truth is we don't have that movie love story because we won't allow it! That's it. That's just the plain and sad reality. It's not some big secret or something out there in the ether. Unfortunately, that's the way it is. We haven't found true love because we're always going against the current. Might as well stop reading this book right now, huh?

Some time ago, I traveled to my hometown with my ex-girlfriend to meet my parents. They invited us out to dinner for the occasion. As the evening progressed, my mom and my girlfriend connected and held a conversation about Hallmark and LMN movies. My girlfriend at the time said:

"Yes, they're so good. I was watching *27 Dresses* some time ago, and there's this part in the movie where Jane says, '*You know how the bride makes her entrance, and everybody turns to look at her? That's when I look at the groom. 'Cause his face says it all, you know? The pure love there.'* So when she's getting married, I couldn't help but tear up seeing how the groom now looked at her completely in love. And I couldn't help but think: Ugh, I wish someone looked at me like that."

As I heard her speak, I remember smiling out of pure, natural social acumen, yet I couldn't help the thought that brought me a brief sensation of sadness: *Well, if you were to only look to your left, I look at you like that.*

It made me understand that as human beings, we are ever oblivious to the affection of others and the importance they give us. It is only natural and perhaps as simple as our biologically written code translating affection into a form of weakness that somehow communicates to us that the person we're dating lacks choices from the opposite sex. If we're someone's only choice, it subconsciously makes us want to look elsewhere for a more "fit" match. Instead of accepting the simple truth: *That person loves us!*

The truth boils down to one of two things. We either settle with partners we learn to love and call ourselves happy while we now and then make "what if" scenarios in our minds about possible future outcomes. Or we're always "learning" from our pasts while telling (lying to) ourselves that *it just didn't work out, that we were just different.* Yet, some external force is always at play. It's never us. All the while, we're secretly feeling utterly alone and wondering, *oh why, oh why.*

I'll let you in on a little secret. We find ourselves in a massive labyrinth of confusion without answers because we are continually going against the current. It's the reason why you're reading this book. In all sincerity, we're afraid of feeling strongly. After all, love *is* a powerful thing, and just as it can have dreamlike moments that put us up in the clouds, it also has painful moments if you happen to slip off from those clouds. So, we create our prison of ideas and excuses never to approach it, reject it when we find it, and then we're baffled and don't know what to do with it when we have it.

That's because love is openness, and openness means possibility and vulnerability. It means the possibility of forging that which we have always imagined into reality and the vulnerability of someone taking advantage of it and hurting us. And nobody wants to feel pain, because—well, pain sucks.

Yet, love exists throughout the animal kingdom, and still, somehow, a bird has more success at following its biology than modern humans. For what purpose? Well, we'll get to it in a minute, but ask yourself what the difference is between how pigs, cats, dogs, and other animals interact and humans, if not superior consciousness? As if life, for some unknown reason, threw two conscious beings into the mix, and the nature of biology was not ever enough again.

The Biology of Love

One time in a past relationship, my partner and I had an intense discussion on the brink of breaking up *again*. Suddenly, the cat jumped onto the bed between us. Completely unaware of our elevated emotions and tension, the little feline only wanted us to pet him. He just wanted some *Luv*. Despite the stress I felt at that moment, I couldn't help but notice and marvel at his poetic innocence.

So many people can relate to this. If your doggo needs some warmth, he/she will simply go cuddle up somewhere, if not with their favorite human. If your friendly feline needs some food, I'm sure you know, you'll hear their meowing followed by their winding between your legs until they get the mission accomplished.

Modern humans acquired a level of consciousness that is not only fascinating but mysterious, as its origin is more or less unknown, and at a degree, that life has never seen in all the eons of time. Yet somehow our success rate in relationships and love is bewildering for lack of a better term—oh wait, found some better ones: baffling, incomprehensible, at the bottom of the chain. In other words, we suck at it.

So how is it possible that Filo and Whiskers have a better chance at love than advanced sapiens? Could it be, again, that our answer is in the question itself? Could it be our consciousness has us overthinking our needs and desires? Is it possible our emotions and all that confusion are something as simple as an evolutionary part of our biology that evolved to continue the beautiful cycle of life? Maybe, but for that, we'll have to take a closer look at the beautiful world of our biology. The tour can be a bit complex, and as they say, *"You can't unscramble a scrambled egg."*

Well, that might be true, but we're still sure as hell going to try.

Consciousness
"Life finds a way."
– Jurassic Park, 1993

Though you could make the point that not all living human beings have necessarily developed this evolutionary trait, we'll focus on the group as a whole for the sake of simplicity, yeah? For us to better understand the biology of love, we first have to understand a bit about consciousness.

The definition of consciousness is *sentience or awareness of internal or external existence. That's nice and all, Hugo, but what does it have to do with love and relationships?* It's coming, so wipe off your Chloe stare and continue reading. The fascinating (a word that doesn't do the subject justice) thing about human consciousness is that life, for some reason, *chose* it to form a part of the evolutionary cycle. Then it formulated a perfect plan to bring it to existence. No one really knows why. For centuries, brilliant individuals have debated, studied, and tried to answer that question and still can't.

However, it is as if life wanted to experiment with this new species we call sapiens by cheating the evolutionary strains of life and skipping the established pattern. And instead of evolving muscles, a robust bone structure, hearing ability, or any of the numerous conventional physical abilities found all through the animal kingdom, life evolved in a way it had never done before (or after, for that matter), resulting in human neurological power, or consciousness. By doing so, humans jumped to the top of the food chain—yet to the bottom of the emotional barrel. However, it needed some potent ingredients to accomplish this feat.

To accomplish consciousness, for one, it needed sapiens to develop through social means for humans to build and grow our brain's thinking and learning abilities through human interaction and stimuli. However, life also needed a secondary ingredient to elicit social interaction in the first place— want to take a guess? I'll let you ponder on that for a minute. But for human beings to develop through social means first, they needed to be born *underdeveloped.*

So, at some point about two hundred thousand years ago, somewhere in the plains of Africa, the first woman with a smaller pelvis was born, forcing her to give birth prematurely and to the great misfortune of all women after her, with a great deal of pain. As we say, the rest is history. Homo sapiens, otherwise better known as modern humans, were born. Yet, if you remember, after all of this, we're still missing a key element to motivate human interaction. Did you guess it? Life needed human beings to **bond**. And so, it embedded in our genes as a bonding mechanism of existence such as *sexuality* and perhaps—*love*.

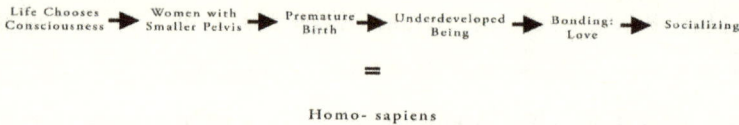

Life Chooses Consciousness → Women with Smaller Pelvis → Premature Birth → Underdeveloped Being → Bonding: Love → Socializing

=

Homo- sapiens

The Psychology of Love

Hey, that's starting to make sense, right? Top of the food chain, eradicate a few species, socializing, spread the seed, evolution, women, bonding, that Darwin guy, a few hours of excruciating pain, and that's it. We have the meaning of love, right?

Well, maybe, but then also maybe Amelia Brand had a point there. We love people who have passed. What part of our evolution programmed us to mourn the dead and feel an emotional connection to them for years after, sometimes until we meet the same fate ourselves? We also love animals and abstract things such as art. What is the evolutionary need in that?

Besides, in the world of human relationships, love can reveal the best and the worst in people. It can do magical, unexplainable things, make us feel one another across distances, perceive the danger loved ones might be in, create art, and fulfill outstanding accomplishments. Love can bring a great deal of happiness, and it can bring a great deal of sorrow. People who once shared so much and loved each other deeply rip each other to pieces, become distant, or deny each other of love once they've gone their separate ways. So what is love's role in all of that? And what does psychology have to say about all this? Perhaps life transcribed something more profound in our biology that can take us closer to understanding this puzzle.

According to Richard Schwartz, a couples therapist, and professor at Harvard Medical School, the positive and negative rollercoaster of powerful emotions we experience with love is associated with two neurological pathways in our brain that communicate with a third essential part of our brain called the nucleus accumbens.

In English, please?

Fair enough. The nucleus accumbens is one of the areas associated with our reward system and dopamine regulation.

Honestly, what does that mean?

It means that two pathways talk to a part of your brain that tells you to get more of something—love, sex, drugs, and more.

The pathways associated with positive emotions connect that reward system part of our brain (nucleus accumbens) with our prefrontal cortex (decision making) while connecting the not-so-positive emotions from the same reward system area to the amygdala (center of emotional behavior). Thus, when we're in love, the system responsible for making critical evaluations of those with whom we are with shuts down. As Schwartz says, "That's the neural basis for the ancient wisdom, *love is blind.*"

If we can learn anything from the above explanation, it's that perhaps we are overthinking ourselves into oblivion. That even though we should be mindful of our decision making, perhaps by being aware of what's going on in our own body and by recognizing our biology, we should trust our nature more.

So maybe after all that, Dr. Brand has a point. Perhaps, it is something we don't yet fully understand, something that the more we dissect, the more questions we have, but isn't that how knowledge works?

As we move along to other chapters, you'll bump into words such as *neurotransmitters, serotonin, dopamine, oxytocin, cortisol,* and *vasopressin,* along with entertaining stories, standard practices, and studies. I'll try to dazzle you with official terms such as *frontal lobe, amygdala, limbic system,* the *ventral tegmental* area, and *prefrontal* this and *hippo* that, and there will be plenty of evidence mind you all. And when you get to the end, I hope the evidence will help divert you from the notion of "true love." A myth forged by you and me unless we decide to be conscientious and recognize all that which we don't know. Maybe it's finally time to bring this home. What do you say?

Our Methodology

The book's entire methodology revolves around four words of action:

1. **Learning**   2. **Understanding**
3. **Accepting**   4. **Applying**

By using four tools to implement and motivate those actions:

1. **Courage** – To face our fear of reality and look inward.
2. **Honesty** – Being realistic and honest with ourselves.
3. **Perspective** – Being open to different perspectives.
4. **Willingness** – To put a conscious effort to move forward.

TWO

Breakups: Why Do I Feel This Way?

LEARNING, UNDERSTANDING & ACCEPTANCE

Nobody said it was easy, but no one ever told you it would be this hard, and it sure as hell need not be, hence the purpose of this chapter. No one ever teaches us to love, and at the same time, no one ever teaches us to cope with breaking up.

Let's face it, breakups simply *suck!* If anything, that's putting it lightly. Breakups can take such a toll on your emotional stability and can affect your personal life in many ways. Your mental health, your work, your self-esteem, and they can permanently impact your personality and who you are as a whole. Now, whether that impact and change are positive or negative is entirely up to you. I hope that the last sentence felt like a breeze of fresh air and gave you exactly that—hope.

It's true, it's up to you. But, it *doesn't* mean you have to do this alone. So strap in and get comfortable because this text is the ultimate guide to get you through the worst of breakups.

If you're reading this as you're going through a breakup, I'm glad you're here, because I wrote this specifically for you. Breakups are confusing times. Your brain is racing, you're sad and miserable, you're full of questions you don't have answers to, and you might be erroneously forming assumptions about your ex. The whole experience can be quite surreal.

As you know, not long before writing this book, I went through a breakup myself. I lost someone I loved very much, when I least expected it, all while facing a centurial pandemic that would be kicking off a quarantine (within four days, to be exact). I know, f*** me, right? I mean, if a bird had shat on me, it would have seemed like part of the breakup package.

As I was going through my breakup, I began a quest to find answers to lessen the pain I was going through. I looked everywhere. You name it— from typing different questions into every search engine I could find, asking close friends for advice, reading books on the subject, asking actual professionals, to even asking different variations of the same question.

I didn't find many answers, and the ones I did find were inconclusive. Despite the answers being right, they said cliché comments such as *It's all a matter of time, it will pass, what you're feeling is entirely normal, and you're not the first person to go through a breakup.* Otherwise, they gave empty pieces of advice such as *you have to accept the fact that the relationship is over; don't dwell on the past, move on.* No s***, Sherlock!

Yet, they failed to tell me what I desperately needed to know—the what, the how, and the why. Why do you say it will pass? How long will it last? How do you know that it will last that long? Why is it normal? Why do I feel like this? What can I do to feel better? How do I move on? How about specific and stronger feelings, such as anxiety and depression? My self-esteem? *Come on, give me some answers,* I thought—all without having my personal Bart Simpson to choke out and physically shake in my desperation.

In all, I understand you, and because I understand you, I feel for you, and with a cliché of my own, please trust that it *will* pass because *understanding* clears the mind and leads to *acceptance*.

In this chapter, you will find lots of specific and detailed advice to answer your questions on emotional, psychological, and biological levels. So, here's to *Learning, Understanding & Acceptance*.

Opening the Breakup Package

When going through a breakup, regardless of the circumstances, you are not going to feel great, and admitting that is the first step to being healthy. It's okay to admit someone hurt you; it is okay to accept your pain. In the first days following the breakup, not much of what anyone tells you is going to make a lot of sense or make you feel better. Your mind is racing; you might be craving to see your ex; you might be in denial, and it's flooding you with a thousand questions. Precisely because of that, let's not ask for the moon and the stars quite yet and take it a step at a time.

For now, keep one thing in mind: *Acceptance!* Our ultimate goal is acceptance. Reaching a state of acceptance will allow us to move on and pursue happiness. But to achieve acceptance, we'll have to use some tools first, implement some strategies, and answer some questions to make sense of the event.

Tools

1. **Paper and pen** – To decipher what happened, for note-taking in this chapter, and for writing down a new routine.
2. **Courage**
3. **Tissues**
4. **Alcohol** (Optional, but highly recommended).

I'm serious—uncross your legs and get these items. I'll wait.

Okay, good, what poison did you choose? Mine is scotch. Anyway, let's proceed.

Breakup Kit

1. **No Fairy Tale Ending** – No contact.
2. **Making Sense of It All** – Defining what happened.
3. **Why Am I Feeling Like This?** – The science behind it.

No Fairy Tale Ending

Some years back, when facing one of those dreadful breakups, I was in so much denial I couldn't keep myself from contacting my ex and trying to get her back. Every time I contacted her, it felt as if I was only pushing her further away. Her demeanor had changed drastically to a point beyond recognition. She proved to be crueler every time, almost as if I was talking to another person. She was at batting practice, and I was the naive baseball that kept shooting at her until she hit a home run. And boy, was she a good hitter.

Two months passed by before I realized I was overdosing on her dose of empowerment, and I decided I'd punished myself enough. After a couple of months of sobering up from zero contact, I received a letter from her in which she expressed she missed me and wondered if I ever thought about her. I didn't respond, yet I realized something more substantial, which led me to a path of research. She wasn't evil. She was simply—*human.*

If you're like my girlfriend, my mother, and many women out there, you might be a sucker for good ol' romance flicks. In the perfect world of romance movies, they sway us away from reality with a happy ending most of the time. You might remember the all-too-familiar airport scene where the male protagonist races off to the airport to catch the love of his life before her plane departs, and she's gone forever. As they see each other, they exchange a few good lines, and after some passionate staring, they kiss just as passionately. Great music further dramatizes the scene, and they live happily ever after. As the credits roll, you sigh, wipe away a few good tears, and then smile, all while the questions pop in your mind: *Could that happen to me? Could that happen in real life?*

I hate to burst your bubble, but the bitter truth is that it *probably won't.* As much as we care for that person we broke up with, as firm a conviction we might have had for the perfect movie moment, something encapsulating the prideful stupidity of humans, and our "consciousness" wouldn't allow it.

What happens in a movie is a simple chain event resulting in a happy ending that looks something like this:

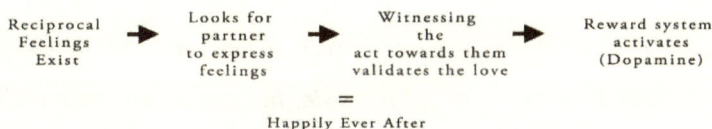

Reciprocal Feelings Exist ➡ Looks for partner to express feelings ➡ Witnessing the act towards them validates the love ➡ Reward system activates (Dopamine)

=

Happily Ever After

In real life, our thinking is a bit more complicated:

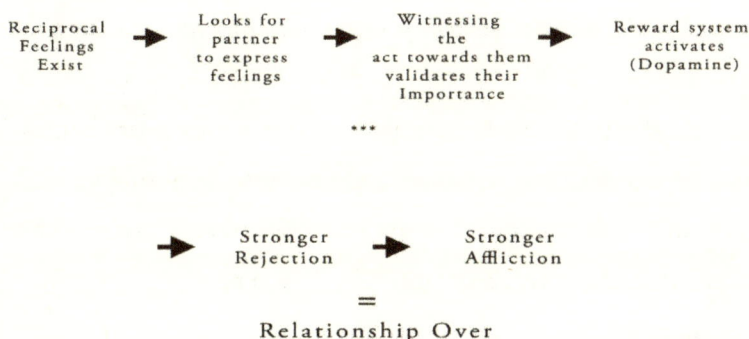

Reciprocal Feelings Exist ➡ Looks for partner to express feelings ➡ Witnessing the act towards them validates their Importance ➡ Reward system activates (Dopamine)

...

➡ Stronger Rejection ➡ Stronger Affliction

=

Relationship Over

Understanding this is simply how human beings work will make it easier for you to access the first tool in your breakup kit—accepting that you *cannot* contact them. Easier said than done, I know, as it sure should be. Every part of your being and every single one of your senses craves a fix of its favorite drug. Your brain is going through chemical changes similar to mourning the death of someone and drug withdrawals. Scary, right?

On the contrary, that information confirms you are not different after all. You should feel relief knowing you are going through something recognized through science to be very challenging. You're going through very *real* withdrawals—relationship withdrawals.

You have to be strong! It's as simple as that. You can't contact them to pursue the relationship, to pick up your favorite pajamas, or to find answers to your many questions, let alone express strong feelings such as missing them. Not only will this decision cause you more pain, but it is a dead-end road with no answers. I promise we'll find answers on our own and elsewhere, but giving in to contacting them will only make you start over, erasing however much progress you've made so far. Let the knowledge itself serve as motivation and repeat: *I feel pain, but contacting them will only increase it. I have to be strong.* **I am strong!**

If Your Ex Contacts You:

Now, if you're on the privileged end of your ex contacting you, how are you to handle the situation? Well, if you read the paragraph above, it's not exactly rocket science, right? It's as simple as act like a mensch; act like a darn human being; don't be a dick. There's absolutely no reason to be hurtful or take advantage of their vulnerability—I mean *zero*. What does it accomplish? What's more, what does it say about you? Why do you need to play on their moment of vulnerability?

Sure, you need to move on yourself and cope with the breakup in your own way, and perhaps you've already been through this phase, and you've weighed your options and made a decision that being together is not the best thing for you. That's completely valid and understandable, but there are a thousand and one ways to go about doing that and removing yourself from the equation. Hurting the person with whom only weeks ago, you shared your life with is never acceptable. Though there're many situations we could use as an example, even in an abusive relationship, it proves nothing, nor do you gain anything. Simply move on.

The truth of the matter is that inflicting pain is also an emotion that's an expression of energy. The literal definition of the word *emotion* comes from the Latin derivative *emotere*, which means energy in motion. So even a hurtful copout such as "I don't feel anything for you" is an expression of your own encaged emotion, one that comes of your own pain and a misplaced sense of feeling better by putting that person down, which is directly related to your own insecurity. Use the power of your imagination and put effort into visualizing how you'd want him to treat you if it was the other way around and do precisely that!

No, but frankly, he has important possessions of mine/we have a business together/we need to talk about substantial issues.

Completely understandable and reasonable. But, unless it's a life-or-death situation, the foundation of your financial livelihood, or something of that gravity, you should completely refrain from contacting them. Also, even in those cases, try finding another solution, or find someone to mediate to deliver such information. Still, even in some of the worst-case scenarios, you can afford to wait. And how long should you wait? I recommend at least three weeks (twenty-one days) to thirty full days of sobering, healing, and coping, maintaining zero contact.

Making Sense of It All

As George Orwell taught us in his masterpiece book *1984*: first there is *learning*, then there is *understanding*, and then there is *acceptance*.

Now, if your partner broke things off with you, the root of all your questions is *why?* Otherwise, if you were the one who ended the relationship, fear may eventually set in, and your first question may become, *Did I make the right choice?* How do you answer these questions and make sense of it all? This part is where you'll need the paper and pen from earlier, and perhaps the drink of choice too. This section is our understanding phase.

It's imperative to understand what you're truly striving to accomplish here. Our purpose is to understand what happened to give you closure, provide meaning, and help you reach acceptance to move on healthily. This exercise is *not* meant for you to dive into the actual *why*. The truth is, the *why* is not a necessary element to our equation. In many cases, the answer could be as simple as that person is just an a*****e. I fear asking *why* would only be counterproductive, making us dwell in an overthinking abyss of questions without answers.

In reality, there could be several reasons why your relationship ended, regardless of who ended the relationship. Some of the most common reasons couples break up are cheating, lack of communication, sex, finances, jealousy, lack of trust, lack of self-knowledge, lack of compromise, disrespect, or psychological, emotional, or physical abuse. Though this news might be disappointing if you were hoping to find a narrowed-down answer, try not to singularize it to your situation and see it as a more ample perspective of life overall.

Otherwise, it could be an endless, painful road leading you to a waiting place as in the Dr. Seuss poem—*Oh the Places You Will Go*. Of course, there are situations where the answer might be clear. To help you, I've narrowed it down to three.

The 3 Decisive Breakups

1. **Cheating**—Deceit, Compulsive Lying, Betrayal

As you know by now, being in a relationship is not easy. It requires work and maturity, among many other things we've described thus far in this book. But as the beginning of the chapter said, it shouldn't be that hard, either. And though there might be extreme reasons people cheat and will go as far as finding them uniquely justifiable, they are *not*. If someone is unhappy with a relationship, they should leave it.

Sure, easier said than done as with most things, but still, these acts should never be allowed, nor compromised. There's plenty of people who *can* change, and ultimately, the decision is yours. But, if you take that risk, and it happens again, just be aware that the lessons in this book are about realization and owning our actions, whether they have positive or negative outcomes. Therefore, despite how unjust unfaithfulness is, remember, *"You do it once, shame on you. You do it twice, shame on me."*

2. **Abuse**—Physical, Psychological, and Emotional

The thing about abuse is it directly impacts your self-esteem. If you face abuse in your relationship, it's *not* a reflection of you but the person inflicting the abuse. Abuse is all about the power of control. The abuser is a procrastinating liar to the mirror; he/she has become a professional manipulator masking an unresolved internal issue of their own by putting on a show of confidence, condescendence, or pure force. This skilled manipulator comes in three forms.

An **emotional abuser** is a living form of passive aggression. They aim to control the emotions that play a part in the relationship. Unfortunately, their passive-aggressiveness can be so ingrained into their personality that they might not even notice they're doing it. In many cases, it has become a programmed self-defense mechanism that forms a part of their everyday life. The most immediate, unofficial diagnosis? They're afraid of the possible conflict that comes with letting go of their emotions.

They won't respond to "I love you," and when they express it, it'll be on their terms. They will act distant and show affection on a limited basis and only when they feel like showing it to you, and will also control the most significant emotional factor in a relationship, *sex*.

The **psychological abuser** will use their words as their weapons. They'll diminish your profession, intelligence, decisions, body—you name it. The hard and degradingly exhausting part of dealing with this abuser is the fact that since they are acting as a bullying motivator, the victim feels constantly conflicted and confused about the contribution this person imposes in their life. "Well, that was smart!" or "That's a nice little project that suits you" are just some examples of how a diminishing compliment may sound. Make no mistake about it, there are unresolved issues within the abuser, and you should pack your bags and hit the road to escape this situation as soon as possible.

The **physical abuser** is the worst and most dangerous of all. This person is dealing with a lot of internal issues that are beyond your help. This person has programmed him/herself to use force caused by an array of different reasons, some being but not limited to: either trauma, mental illness, learned personal experiences, etc. They will often send themselves on a guilt trip after their abuse in which they'll act like the sweetest human on earth, claiming their unconditional love and swearing not ever to do it again. Oh, and they'll have the oblivious nerve to blame you for upsetting them. But how can love ever directly inflict harm on you? Seek courage from friends and loved ones, don't be afraid to speak out, call the police, and seek professional help. Know that there are many things you can do to get out of this relationship as soon as possible.

3. Unable to Compromise

It's often the case that couples did have great chemistry, loved each other, liked each other, and formed great memories together, but at the end of the day, they were unable to compromise. We'll discuss compromise more thoroughly in chapter thirteen.

In other situations, it could be a compilation of many minute different things that could not be solved, or perhaps solving those issues will take more effort than either one is willing to exert. But if you've addressed the problems, you've worked on them proactively, and after a time you can't reach a compromise, breaking up might be for the best.

Sometimes, people just wrong us. Sometimes, people just don't see our worth. Sometimes, people don't have the emotional insight or maturity to be in a relationship, and sometimes *it is* on us. So, let's not build an adamant expectation for finding a concrete answer, as one might simply not exist; some things are just not in our control.

A better question to ask that will help us define what happened and make sense of our situation in a healthy manner is:

What determining factor wasn't working in the relationship from which I can learn?

Discovering a Learning Flaw

Asking you to summarize a series of experiences and package them neatly into a broad definition is no easy task. The process can be taxing and draining, yet the sooner we find this answer, the more comfortably we'll reach a state of acceptance. If you find this process is detrimental to your current state, please don't hesitate to move on to the next section. This exercise is not to hinder your progress but to help you. For now, writing down the question is a big step in your progress, and you should be proud of your courage.

But how do I find an overall key factor from which I can learn?

I'll give you a couple of examples and break it down into a simple, two-part process you can follow to help you draw this out. If you feel confident doing this on your own, feel free to jump to the next section: *Why Am I Feeling Like This?*

Ownership: What could you have improved? Was it acceptance? Trust? Communication? Compromise? Or, all of the above?

We hate to admit this—it might even sound insensitive—but we need to face reality. If the relationship ended, something was failing, whether this was by your account, theirs, or both. Sure, there are occasions where the breakup might have taken you by utter surprise. Maybe you had just gone on a date, discussed an upcoming trip, or planned the weekend with friends. Perhaps you were that perfect girl, and your partner didn't appreciate you. But in that case, what could be more of a compelling answer than knowing someone didn't appreciate you? Become the one who got away!

How do I apply this positively to my life? The irony is that the selfishness of internal growth and self-improvement affects our future relationships. As Tony Robbins tells us, we must fix ourselves before we take ourselves to the next relationship and ruin that one because we didn't take the time to know, work on, and fall in love with ourselves first.

"Loving yourself" may sound like a cheap cliché. It might be if there was no process of learning and didn't allow the passing of time for healing and applied a conscious effort for improvement.

Finding the issue and transforming it into a lesson is the real gift to yourself. A gift that will allow you to cope with the upcoming weeks and enable you to heal, grow, and fall in love with *you.*

Adopt a winning mentality—you didn't lose, you'll learn. In the world of business, for example, we are continually facing adversity, and regardless of how much effort and passion you put into a specific entrepreneurial journey, sometimes it simply doesn't work because of competition, the market, pandemics, legalities, etc. Suppose we saw a lack of cash flow or a business bankruptcy as a failure, then we wouldn't survive. We must have a learning vs. failing mentality to move forward and succeed.

If someone left you like a tree in the wasteland, show them the forest. A baboon can't differentiate a diamond from a garnet if he stumbled upon a stone. He might try to pick his teeth with it, so try to see the brighter side of things and try not to dwell on it. Read this chapter, go through the motions, and give that diamond to someone who sees its shine.

On the other hand, if it was you picking your teeth with the diamond, well, be honest with yourself. Otherwise, you'll go through life without ever finding that special someone, wondering why it's not working out. Invest time in really getting to know yourself. Were you co-dependent? Were you too agreeable or compliant? I'll admit that in a past relationship, I became the "yes man" that is so popularly unattractive. Your answer lies in the issue; just be honest and fair, but don't put yourself down. You're human, and you're going to be making mistakes. There's always room for improvement. Now, you must take action.

Why Am I Feeling This Way?

Loss of appetite? Lack of energy? Do you feel drained or exhausted? Trouble sleeping? Anxiety? Depression? Nightmares? Cold sweats? Is your mind racing and unable to focus? Unshakable thoughts of your ex? Low self-esteem? Do you feel sick? You may be experiencing these things all while you've been staying strong, not contacting your ex, and keeping distracted, ultimately leading you to the question, *why am I feeling this way?*

If you happen to confess your poor state to your close friends or family, you might find yourself a bit discouraged hearing the all-too-common cliché, *"It's completely normal."*

Okay, sure, but what's next? Why is it normal? Don't waste your time asking others for the answer. It's a dead-end road that will leave you more flustered than when you started. Don't get me wrong, your real friends mean well and have your best interests at heart, yet because they love you, they don't want to hurt you and can often find themselves at a loss for words when trying to offer comfort.

The Cocktail of Love:
DOPAMINE, OXYTOCIN & SEROTONIN

Picture a lab inside your brain, or otherwise, imagining Walter White cooking up a nifty cocktail in your head will have to do.

The precise function and purpose of these brain chemicals, as well as the areas of the brain where they play their symphony, are complicated, to say the least. So many aspects of their mere existence are not only still unknown but remain a subject of controversy throughout history and debate among psychologists and professionals today. I promise not to make your head explode as we simply learn enough to understand and set our minds at ease.

Dopamine, oxytocin, and serotonin are neurotransmitters or hormones depending on their function and area of the brain in which they operate. In layman terms, a neurotransmitter is no more than a group of chemical agents released by neurons to stimulate impulses to other neurons (nerve cells) throughout the nervous system. When these fire a couple of billion times per second, you get a message; you might know this message as what we call memory, an image, a smell, an emotion, or physical feeling.

Dopamine: This hormone alone has about 110,000 research papers to its name, so as you may imagine, its function is extensive and complex. We will specifically focus on VTA (ventral tegmental area) dopamine neurons, which activate the reward system in our brain. These dopamine neurons trigger when something unexpectedly good happens, telling us to get more of the same, like food or drugs. Fun stuff, huh? Well, sometimes. In our case, these neurons activate by the presence of our significant other, perhaps written into our biology by the author Life as a mere impulse to mate and reproduce.

In essence, when experiencing the presence, affection, or one of the many different factors associated with our significant other, the dopamine in our brain is telling us to get more of *them*, hence why you crave seeing them so much.

Oxytocin: If you have ever been in a hospital room after a mother has given birth, you might have noticed a solution by her side labeled "oxytocin." Though the purpose in that scenario is to prevent hemorrhaging, it still serves as a reference.

Oxytocin is more commonly known as the "love hormone." It's primarily responsible for bonding, which is one of the main pillars of our human existence.

Sapiens is the only species ever to evolve consciousness, which was achieved by premature birth. The reason for a premature birth was to elicit a slower development: bonding. In the cocktail of love, oxytocin plays a role in our social interaction and is released when we hug, kiss, and have sex, influencing trust, empathy, and even orgasms. This explanation is why the hormone is such a critical factor in couple bonding.

Serotonin: As with the chicken and the egg scenario, whether serotonin influences dopamine or dopamine rewards us with serotonin is not a scientific certainty. In its most straightforward fashion, what we do know is that serotonin is widely known as the "happy chemical." Like most scientific terms and its neurotransmitter companions, it is responsible for a wide range of functions. Still, for our understanding, serotonin is directly correlated to our happiness and regulating our moods.

Those are only the main ingredients in this potent cocktail of love, cortisol and vasopressin also join the party. Cortisol is associated with pain and stress, while vasopressin binds oxytocin to lead the bonding mechanism (which is in overdrive during a breakup.)

Now, if I offered you a cocktail of such description, you would probably throw it in my face, or take off running, followed by serving me a hefty lawsuit. Don't worry, it's on the house, and you don't have much say in the matter since they're neatly working in your brain in this very moment; hopefully, mainly serotonin as you're reading this book. Consequently, when you're going through a breakup, your brain lacks its accustomed regular supply of these neurotransmitters, sending you into a neurological and painful withdrawal.

No other word combination could be more literal than a *painful withdrawal*. Studies have shown your brain processes a breakup the same way it does physical pain. As well, a breakup's neurological effects are on par with some drug addiction withdrawals.

A study by Columbia University tested participants who went through a breakup during the previous six months. Each participant completed two tasks while undergoing an fMRI (functional Magnetic Resonance Imaging). One assignment involved feelings of rejection and another physical pain.

For the first task, the researchers showed the participants an image of their ex-partner and asked them how they felt about the breakup experience. On the second task, the researchers attached a thermal stimulation device to the left forearm of the participants. Researchers then compared the scans from when the participants looked at their friend's picture to the physical pain caused by the hot probe on their arm.

The researchers found the same areas of the brain lit up when the participants saw a picture of their ex and experienced pain through the probe. These regions include the insula and cingulate cortex, which play an essential role in pain processing by sending output signals to the amygdala, an area in the brain mostly known as the processing center for emotions.

So, if learning leads to understanding, and understanding to acceptance, then this knowledge should further confirm not only how normal you are for hurting, but enforce the beauty of what a marvelous creation you are, ultimately dismissing the notion that you are *alone.* Because as you now know, it's the opposite; your brain is busy working to help you process the experience. Your mind continues to work towards a brighter tomorrow. So, don't give up, do as it does—*get busy!*

Now, given all the information above, connecting the dots and understanding the relationship withdrawal you might be experiencing might be common sense. On the other hand, its very nature might still be puzzling, making you wonder if you're cuckoo by drawing comparisons to actual drug addiction withdrawals. The good news is you're not cuckoo, as studies have shown similarities in both. After all, it does seem like a bit of common sense, doesn't it? You're having intense cravings to be with that person, to see them, or at least know about them. These cravings present themselves as constant images of your ex flooding your mind at what seems like the speed of light, making it hard to focus on your surroundings by continuously pulling you into your mind.

A study by researchers Helen Fisher and Helen Brown found different areas of the brain, including the ventral tegmental area, the nucleus accumbens, and ventral striatum, which are part of the reward/motivation system, communicate through the discussed neurotransmitter dopamine. That very same reward system is what's primarily associated with drug addiction. The natural depletion of dopamine after your breakup will give you genuine addiction withdrawal symptoms. Other studies even suggest with more specificity that love withdrawals are on par with cocaine addiction withdrawals.

So, remember that first paragraph in this section? The anxiety, the unshakable images of your ex, the cravings to know about them, see them, contact them, the fear? Well, now you can at least understand why, which takes us closer to understand how to cope with the "delightful" experience. Your depleted brain is desperate to replenish these chemicals by *any* means necessary, so it's flooding you with every memory of your ex it can find.

But is that all we are? Merely a recipe made of some chemical ingredients thrown into a boiling stew? As systemic and bane as it may sound, to an extent, yes, at least as far our emotions, feelings, and internal experiences go.

Maybe after a couple of minutes spent reading these paragraphs, a now empty glass of wine, and hopefully not too bad of a headache, you can breathe a little easier knowing you are far from *weird* and be a bit more at ease that it will most certainly pass. And no, you are not going insane—it just feels that way.

THREE

Breakups: Coping 101
THE COMPREHENSIVE GUIDE TO COPING

Upon hearing the subject of psychology, the first name that might pop to mind may be the all too famous Sigmund Freud. Yet in modern psychology, the poor lad doesn't get as much love as he used to. Though his merits might have been the seed that allowed the tree to grow, his method of psychoanalysis is no longer recognized as a convenient approach to help patients and is now rarely practiced.

CBT or Cognitive Behavioral Therapy has become the new popular kid in school due to its more direct approach and proven rapid results on a high percentage of patients that undergo therapy. Cool, but what is it? Cognitive therapy helps patients develop alternative ways of behaving and approaches to thinking, which may help reduce psychological distress.

What good is understanding all that information and reaching acceptance if you don't have some guidelines or tools to implement a strategy? In the name of CBT, cheers to some behavioral approaches.

Men & Women Cope Differently

*"Lisa, why don't you just go. Lisa, seriously. Lisa, we're here 'cause
we're concerned about you, and he's a dickwad. We want you to just go."
"I can't just go. I can't just go, Karen. It's not that easy, okay? Like you
guys think I can just go. It's not that simple, all right? My CDs are in his
truck."*
– 2006, *Vicious Circle*, Dane Cook

After listening to that standup of the always super-funny Dane Cook, I couldn't help but notice something beyond the context of his words. It hit me at that moment—women are more supportive of one another.

Statistically, women and men cope differently. Women usually mourn more profoundly and often face more substantial withdrawals but tend to get over the breakup faster than men. Regretfully, this fact doesn't have as much to do with a psychological advantage or chemical difference between the sexes, but more with the implied socially accepted biased telling men to "suck it up."

In contrast to men, when you're going through a breakup, it is more likely for girlfriends to motivate you. They'll call or stubbornly come over despite you wanting to sob in your room by yourself; they'll trash-talk your ex for you (thank you), and their visits will usually be in the company of a bottle of wine for you to vent freely. If anything, they are genuinely interested in hearing your story. They will listen to you for as long as it takes as they console you with hugs, and Tiffany hands you tissues to wipe the tears and—well, the rest.

After you've vented enough and your girls have been able to get a few good laughs out of you, and only a vestige of tears are in your bloodshot eyes and on your red face, they'll distract you by changing the subject to something more benign. Though your grieving has just begun, at least for that moment, your friends remind you they're there for you through thick and thin, and that despite what you feel, you are not alone. What could be a more poetically beautiful image?

Now, for men, it's a tad different. First of all, when a man goes through a breakup and meets his best friends to vent about his present predicament, he'll majorly dilute the story. At the same time, making sure to leave out some of the hurtful parts due to some senseless testosterone-driven pride. Secondly, when he finishes his narration, he'll be met with: "Yeah, that sucks man." If you were waiting for something else, please don't hold your breath. Okay, that's a bit harsh. Male friends will most definitely lend a guy more straightforward advice and non-diluted support, and what's more, they'll give him that much-appreciated and needed male perspective.

It's not a competition. No one is to blame. The point is that despite facing social norms, there's no written rule saying we have to abide by them. Calling a spade a spade, men need women friends that will be more open to listening and comforting them, as much as women need the more rigid opinion of men. Supporting you out of pure love and empathy can easily lead to blind cheerleading without giving you a dose of the challenge you might need.

As I went through my not-so-smooth breakup, I confided in my male friends for motivation, a pint of their brutal reality, and some testosterone-filled advice. Yet, my venting and emotional support mainly came from beautiful women who were there for me through some of the most challenging moments.

Veni, Vidi, Vici

I came—I saw—I conquered. As you learn, understand, and accept, now, like Julius Caesar (or Cleopatra), you must conquer, and conquer we shall. Make no mistake about it, and to clear things up for once and for all—this is a *fight!* One, you are going to win because you are strong, you are important, you are a warrior, you are a goddess!

As you're suffering through the symptoms as mentioned earlier, it feels like your system is attacking you, so you must strike back. With that in mind, motivation and energy may not be at your disposal at the beginning of your breakup, so creating a routine and forcing yourself to abide by it is crucial. Come on, grab your paper and pen.

Our strategy is categorized into two different components/sections: Fundamental Actions and Recommended Coping Activities

Fundamental Actions are chemical regulators, the steps that, upon taking, will encapsulate a great deal of your overall wellbeing. Keep in mind they are all just recommendations, of course, yet deviating from our plan might halt your speedy recovery process.

Recommended Coping Activities are the exercises, strategies, and techniques that complement the former. They are optional processes for memory manipulation, relaxation, and structure for our next chapter.

Remember, your natural chemical levels are low, so we must replenish them by natural means. And no, intimacy, dating, and jumping into another relationship are all but natural, as you'll learn in sections ahead.

Fundamental Actions

1. **Sever all contact**

Making this decision is imperative for anything else in these chapters to work, and getting these chapters to work is a direct benefit to your health and giving the gift of you to a future special lucky someone. So take your time and remember this is not a race. It's okay to take it slowly, and it's okay not to feel greatly optimistic—as long as you fight and push forward. But giving up and giving in are not ingredients in these chapters. Failure is not an option!

Do me a favor. All I ask is you simply keep two small things in your mind that will arm you better to fight off and ease these urges: timeframe and consequence. Given the scientific explanations of your chemically charged brain in our previous chapter, keeping a magical timeframe of three weeks will at least help you have an optimistic expectation for the symptoms to decrease if not dissipate dramatically. Lastly, putting consequences into perspective should help you see what awaits if you break your no-contact streak—*relapse.* If you don't like your feelings today, you sure as hell won't enjoy how you'll feel tomorrow if you contact them. Like any other addiction, this response will reset your clock to day one.

2.   Dispose of all mementos

Whether you lived together or not, it is natural you'll have "things" or personal belongings of your ex, whether it be clothing, books, ticket stubs, you name it. Regardless of how small an object is, you must remove *all* of these items from your sight. No excuses, and though you can surely throw them away, you can also simply store them out of sight, as well as giving them all back (later).

Your mind needs no extra stimulus to bring up the memories of your ex. Its hands will be quite full as it is. For the next couple of days, it will overwhelm you with a storm of memories all on its own, don't worry. So be kind to the person in the mirror and help her by removing every single reminder and creating a clean place of Zen. Don't be slick; this includes your phone storage and social media. Mark my words, once you follow through with this task, you'll be able to breathe a little better.

3.   Exercise

Remember that nifty, heady little cocktail of love? Well, let me introduce you to your new best friend—endorphins! These are neurotransmitters with the muscle to bully away some of that stress, pain, and fear. See them as your own handy, free-of-charge, opiate-like narcotic. There are many types of endorphins, and they primarily play a role in the function of the central nervous system by repressing or motivating the signals of nearby neurons. When you exercise, your body recognizes the act as a strenuous activity and triggers the release of this beautiful hormone.

A second thing happens when you exercise. Not only are you bringing endorphins to regulate traffic into the picture, but you're also naturally releasing energy in the healthiest of ways. If you recall from our previous chapter, emotion is energy in motion. It's no secret you're dealing with a bit more emotion than what you're accustomed to in your everyday life. So instead of letting those emotions run loose inside you creating havoc, open the door, letting some of those a*****es out. Bottling those emotions up will only serve to destroy the masterpiece that is you.

Don't feel the energy to work out? Force yourself to help yourself. All exercise is good exercise, we all have days where we just can't face half an hour of HIIT, let alone an hour of strength and conditioning or a kickboxing class, so try low-intensity exercises like walking, stretching, yoga. To fall in love with yourself, you must start with acts of kindness. You have the best interests at heart for the person in the mirror. So, embrace the warrior in you, get off your rear, and let's get to this workout.

Jogging? Cycling? HIIT? Jump-roping? Free weights? Something else?

Commit:

---

4.   Crying

Cry, cry some more, and cry again. The research and benefits of crying are extensive and heavily debated. There's an array of implications to be made about the possible adverse effects of crying, depending on an ample number of personal situations. I think crying in moments of mourning can be extremely beneficial as it also releases energy and hormones associated with stress, such as this mouthful: **adrenocorticotropic** hormone. This hormone happens to be directly correlated to the limbic system of the brain; hence, stress and pain.

Besides, why not own our tears? As far as we know, crying is a uniquely human function. Dr. Lauren Bylsma from the University of Pittsburgh suggests the second function of emotional tears might ascribe to our biology. Meaning that being the social and bonding animals we are, as an evolutionary trait, we developed crying in front of others to cause empathy, which in turn can encourage people to comfort us. In all, don't hold back, and whether it is in front of a friend or the company of just you, let yourself go and don't self-criticize. Your body will know when enough is enough.

5.   Expression

Expression is a critical factor for understanding your situation, expressing your emotions, and being human. That's right, the ability to express is written into your biology, and we have the human privilege of options—text yourself (I do this myself; quite therapeutic), keep a journal, phone a friend, or speak to the wind. Finding a method of expression that works for you is crucial for a healthy recovery. Expressing is so important in different stages of human life that we've dedicated a whole chapter to it in this book, as you'll find out ahead.

Expressing ourselves is not always easy. It can often make us feel vulnerable, and the reason is quite simple—we *are* vulnerable. After all, we are exposing an intimate thought, experience, or feeling. Naturally, you might feel a bit wary or awkward at first, but accepting your vulnerability is in itself an act of courage, whispering to you that it's okay. If that is not entirely convincing yet, understand this scientific fact: your DNA, your brain are *not* wired for isolation. They simply are *not!* It didn't work for Newton or Tesla, and rest assured, it won't work for you. See this as a necessary action as it is imperative for your wellbeing. Failing or not doing a great job at our attempts to express is completely normal, as you are facing a genuine crisis, do it at your own pace, but express away.

What will be your method/s of expressing? Fill in the blank:

_____

6. **Food**

If there was any time in your life to indulge, this is as good of an excuse as you'll ever get, so as long as it is with moderation, this might be the time to treat yourself to an occasional favorite treat! This moment might be the time to treat yourself to that delicious slice of chocolate cake you've been drooling over. Let the occasional carbohydrate light up your brain like a Christmas tree and give you a slight boost of the serotonin you've been needing. In other words, it stabilizes your mood and increases your happiness.

Remember the ingredients to that potent cocktail? Who can forget, right? One of those essential ingredients was dopamine. Well, Dr. John D. Fernstrom from the Department of Psychiatry at the University of Pittsburgh School of Medicine explains that increasing our protein intake stimulates dopamine production. An increase in dopamine means quenching some of those passion withdrawals we discussed. The specific amino acids responsible for this chemical synthesis are tyrosine and phenylalanine, which are in foods such as beef, turkey, eggs, dairy, and soy. So, either look you up a tasty recipe or perhaps it's time to sign up for that cooking class you've been putting off. Regardless, protein it up!

If you're like me and many others, you might be lacking an appetite, and that's perfectly normal as well, but when we lack the energy or motivation to eat, our decision must take over. Remembering that food is the vital source of fuel for our brain is essential to take ourselves to the kitchen and reverse engineer the process.

Remember who this is all about—you. Be kind and treat her to a deli snack as it may very well be the energy she's been needing all along.

Insert Favorite Snack/Dessert:

---

7.   Music

Music affecting your mood, emotions, and wellbeing might not be news to you. You didn't have to read a detailed neuropsychological study by a reputable university to come to that conclusion yourself.

Letting the mystic chords of soundwaves soothe you with their caressing touch might be a phenomenal idea during this time. I have two easy suggestions that could prove useful if you implement them in your strategy. One, make a specific playlist of songs with a one hundred percent success rate on your "happy" scale. Simple enough, right?

Secondly, I have a relaxing exercise for you. Put on your headphones, open your music app, and listen to music for a day or two, searching either by a soothing or energetic genre, such as classical, opera, or electronic music.

On the other hand, you could gauge purely by emotion, asking yourself, *Does this song make me feel good?* If you answer yes, add it to your playlist. Then, you'll have a go-to playlist guaranteed to send you to the right place in your mind for the following weeks. I created a playlist[1] with over two hundred songs. Try it out, and let's see how many songs you end up.

Now, if you are a "facts" person, the McGill University's psychology department led by Professor Daniel J. Levitin did a meta-analysis of four hundred music research studies. The study found that listening to music can reduce stress and fight depression by directly lowering levels of cortisol in your system. Another study by the research team at the Center for Interdisciplinary Music Research[2] in Finland organized music into different categories and found exciting results. After testing participants through fMRI scanning, the researchers found that women who used a **diversion method**[3] had increased activity in the area of the brain associated with decision making and regulating emotions called the mPFC (medial prefrontal cortex). This part of your brain might be broadly associated with your superpower of human intelligence.

8.   **Sleep**

Through one means or another, we've all heard about the importance of sleep. Yet, regardless of repetitive clichés, they sometimes have a valid point; rest happens to be one of those. As you've been reading, it might seem the

---

[1] You can listen to the author's playlists by visiting his profile on Spotify https://open.spotify.com/user/125091040?si=nnsotMJZSjqYqz8U6A05Mw

[2] Center for Interdisciplinary Music Research is a research facility at the University of Jyväskylä, the University of Helsinki, and Aalto University in Finland

[3] Distraction; music opposite of your mood.

same vital players keep coming up—serotonin, dopamine, oxytocin, cortisol, chemicals, and more chemicals. And since the intention is not making the problem worse by giving you a headache of your own, as you may suspect, sleep will help regulate these protagonist hormones.

Ironically, and in contrast to the importance of sleep, getting a full night's rest might be proving difficult during the process of your breakup. So let's focus on applying practical strategies to build a sleeping routine that can make it easier for you to get eight hours of sleep:

- **Turn down your thermostat.** Either you've already been doing this and are not quite sure why it works, or you might find yourself saying, "*What? I'm cold all the time. How does this make sense?*" Well, your body naturally drops one to two degrees when sleeping. Experts recommend keeping a cool ambiance while sleeping for your body temperature to remain at a steady, desired temperature. The optimal thermostat setting is 69°. Snuggle up.

- **Take a warm bath or shower before heading to bed.** All the same, taking a warm bath before going to bed causes thermoregulators such as sweat to cool you down faster. Ultimately, it causes you to get to your desired temperature more naturally and with ease.

- **Tea**. A good, ol' natural herbal solution to insomnia or plain having trouble sleeping may be as easy as a nice hot cup of tea. According to Kaitlyn Berkheiser, a registered dietitian writing for Healthline.com, herbal teas have been a reliable solution for sleeping problems for centuries. Now backed by medical data, Kaitlyn recommends the following for a comfortable, restful night: chamomile, valerian root, lavender, lemon balm, passionflower, or magnolia bark. Take your pick!

- **Relaxing activity**. TVs, smartphones, and screen monitors are not necessarily the friendliest of methods to set you up for a good night's rest. All these devices emit blue light, repressing your melatonin production, which is responsible for making you sleepy at night. Cell phones and its applications are designed to keep you hyper-focused, which can cause you to scroll endlessly, despite having a blue light setting. Reading a book while you lie in bed or writing your thoughts in a journal for tomorrow are great and healthy substitutions. Your Nyctophilia, Soothing, Hot Choice:

Recommended Coping Activities

1.   Mindfulness Meditation

You are a pattern-recognizing machine! Whether you were aware of it or not is a different point, but you are a processor taking in information regularly. Your senses are so sensitive you don't even notice them more than half the time. Sometimes, it's time to reel things in for a moment and become aligned with those same senses. See it as a slight wheel alignment that was long overdue. If you haven't tried meditation yet, it might be because the task itself sounds like a tedious, boring handful of worthless baloney. And I can't blame you. Society is always pulling us in different directions, and social media has made us addicted to its apps, reducing our attention spans to where having a straight conversation has become a daunting task. It's no wonder we can't keep out of our minds and sometimes feel like we haven't seen our present surroundings.

Though meditation could be an action highly beneficial on a daily basis, asking you to take on a new task seems like a bit much. Instead, learning a bit about meditation to add as a distraction option to your toolbox seems like a better idea.

Mindfulness meditation is the act of focusing back on the present for a period no longer than ten minutes. It teaches you to inhale and exhale slowly while concentrating on the stillness in between your breaths. It is the act of focusing on each of your five senses, becoming completely aware of your surroundings. Noticing details such as smelling the damp room, feeling the texture of your clothes and your weight on your buttocks, seeing the surrounding ambiance, noticing its colors, hearing the loudness of silence, ever more acute to the chirping of the birds outside.

If you're interested in giving it a try, visit my page HugoBradford.com that has some great options that will guide you through the fastest ten minutes of your life.

Breathing has a direct correlation to our ancestors. In harsher times, when hunter-gatherers roamed Mother Earth, and comfy memory-foam beds and fancy apartments weren't an option, our ancestors faced many dangers. Upon facing these dangers, their bodies would activate a fight-or-flight response, and our ancestors got to run, and run they did. If they were successful in escaping their hungry predator, what would you guess would be the first natural response they had? Exactly—gasping for air, inhaling and exhaling deeply.

After thousands of years, the action of breathing in deeply registered in our brains as a sign telling us we're safe. Whenever you practice inhaling deeply and exhaling slowly, you are telling your mind you are safe, which in turn helps you relax.

2. Activities and hobbies

Often, a breakup can leave you so scarred that you question your identity. You are right in asking this question. You are in the process of finding yourself, and since you're reading this text, it is safe to say you've already taken a step in the right direction towards helping yourself.

The easiest way to approach all problems is by asking the most superficial of all questions. In our case, that question is: *What do I enjoy doing by myself?* For instance, I enjoy biking. In my relationship, my ex-partner never wanted to go on bike rides, despite me buying her a bike, or her having one before. Be that as it may, when my relationship ended, it hadn't been twenty-four hours when I was on my bike riding through the beautiful green landscape that makes Bayou Park the attraction it is.

Give it a shot. Focus on the solution, not the problem, which is what we tend to focus on, making it bigger than it seems. I assure you, there was something you enjoyed doing before you met your ex, something you wanted to do, and your partner didn't; or simply something new you wanted to try, the thing you think about and haven't gotten around to doing. But if you do, you'll find yourself smiling.

The first time I rode my bike after my breakup, I remember thinking, *Gosh, it reminds me of her, I miss her,* despite being with my pretty friend in a beautiful park. Yet after a couple of seconds, I couldn't help but laugh at my thought. *Wait, what? Why do you miss her? You never rode with her, silly!* I burst out laughing, and my friend asked me why. You'll find you have the power to surprise yourself. Give it a go!

3. Hang out with friends

We previously capitalized on the importance of having our friends around in times of need, such as this one. So, for the sake of avoiding redundancy, we'll merely highlight how to better benefit from their company.

It's crucial to challenge yourself in your solitude and be conscious about not simply replacing dopamine with another source of dopamine—in other words, substituting missed company with *other* company. On the contrary, during the first weeks of vulnerability, do not hold back from the company of great friends. Who am I to tell you otherwise? I had my best friend and business partner visit me from Austin three times in about five weeks. We focused on work the majority of the time. He never told me that was the reason he was visiting, but I know why he did it.

Remember to mix it up. You need both your girlfriends and guy friends. Just like I stuck with Mari, Sebastian, Sarah, and Sketch, expressing to both Venus and Mars will give you two very different routes of communication, you would have limited yourself from otherwise. Women tend to be more comforting. They take some of the weight off you by grieving with you, they'll bad talk your ex and voice the insults you didn't allow yourself to say, and as I said, they'll never fail to bring the alcohol.

Meanwhile, men can be a bit sturdier, and while not as emotionally comforting, they serve you a much-needed unsweetened tea with a bitter taste of reality. They'll be more honest with you about the true intentions of men (since they might have those intentions themselves...*cough, cough*). The point being, don't be afraid to seek out your friends with the thought that you might be tiring them out. Because the truth is you might, but the more important fact is it doesn't matter. Your best friends will be there for you regardless of how tired they are of our whining self. So have a good laugh and call them up, because a time may come when you need to return the favor. What are friends for, right?

4. Flirt

Take this point with a word of caution: be careful of this leading to sex, intimacy, or dating. For the sake of explanation, if we put this point back into the context of drug addiction withdrawals, people going through withdrawals are commonly given drugs by professionals to hinder the intensity of withdrawal effects. In the name of hindering, doing some Tindering might not be entirely out of the question—as long as we are conscious of our control or lack thereof, and more importantly, our purpose. Yet, if you don't feel like flirting nor are you craving any attention, great, that's fine, continue on your path of self-love.

The purpose is no more than getting the mere and harmless attention of people we find attractive by flirting for a couple of weeks until our symptoms subside; no more than a healthy, small dose of our favorite drug to ease the symptoms.

If you find yourself feeling seduced by a total Don Juan and his alpha, hypnotic charm— STOP!

Darn! Am I going to die for having some good, ol' casual sex? What's the harm, Hugo?

No, you're not going to die from exposure to some male anatomy, Susan. But this text is not about human anatomy, it's *The Anatomy of Love*, and our overall goal is healing, and the general purpose is finding a healthy relationship. To do that, we must discover, heal, rebuild, and love ourselves first.

5.   Goal setting

Are you familiar with the quote, "A goal without a plan is just a wish" by Antoine de Saint-Exupéry? We've already established the purpose of our immediate course of action is to heal. But what happens after that? How do we move forward onto both the literal next chapter and the next chapter of our lives? How do we make the leap from *"Hooray, I made it, I feel great, give me my graduation diploma"* to a sober, conscious *"I'm ready! I've healed, found myself, and am ready to be found"?* By goal setting. By putting the view of life into perspective and assigning ourselves an idea, an image, a task that will make us perform a set of responsibilities to get from point A to point B and break the tape across the finish line.

Is your goal to make the earth shake upon the wrath of your success? A diploma? Postgraduate education? A new body?—Or perhaps unleashing your mind with ink and paper to write an entire book in the span of a few months. Regardless of what path you take, *take a path*. Please do not wait for the course to reveal itself; go, and create it. Life will pass you if you wait for life to happen, so be proactive and dictate your life. Or in the great words of motivational speaker Eric Thomas:

"You're already in pain. You're already hurt. Get a reward from it!"

Memories vs. Imagination

What is the difference between a memory and envisioning the future? I will let you ponder that for a bit. When seeking the advice of friends or close ones about coping with your breakup, you might perceive their tone as insensitive, and that's probably because *they are* a bit insensitive to it. Cut them some slack. They probably mean well. It's not personal, and it's likely unintentional. With time, the healer of all things, they have become unattached from the emotional experience of a breakup. Our brain does this naturally; it hyper-focuses on the problem and learns from it, and then it fades. If anything, see that as hopeful proof that you'll be like them in a matter of time.

People naturally forget the grueling nature of the process, but when putting it into perspective, you're dealing with a lot. Not only do you have to deal with all your breakup symptoms, but also with removing your memories while expunging the images of the future—the trips, the promises, the house, making a family. *It's not easy.*

Write your answer to the question above:

_____

If you wrote anything along the lines of *nothing* or *not much*, you answered correctly. Usually, when I ask people that question in person, I often get something like:

"Well, my memories are my reality, and my thoughts of the future are my hopes."

"Great. What makes your memories real as opposed to your thoughts of the future?"

See the point I'm getting at? The difference between a memory and your thoughts of the future is—*belief.* When you engage your surrounding by looking at the room around you, what physical power does a memory have? None. Remember Newton's first law of thermodynamics: Energy can neither be created nor destroyed. The power memory has on your present is all a matter of choice. The aloof, sensible truth is that our doubts, insecurities, and self-esteem are all self-inflicted.

Not only does this serve as a revealing and practical philosophy, but it also holds in science. The areas of the brain associated with imagination are also the areas of your brain linked to memory retention and learning. Taking this into account, if we reverse engineer this process and use it against itself, consciously using belief as the means of programming, what is there to say it wouldn't work?

Welcome to the conscious programming of your mind by teaching you how to use your uniquely human power of imagination to replace memories and gain control of those fuddled emotions. Therefore, let's engage in another fun task.

Focus on substituting three vital images encapsulating your memories into branches upon branches of torturing moments.

1. Person

I'll choose this one for you—your ex. The image of your ex is the epicenter of your entire present unpleasant experience. So, write "Point One" and then a paragraph describing the partner you want in your life with as much detail as possible. How do you imagine he looks? What are his qualities? What are you both doing? Are you laughing? Are you drinking or being lazy? Let your imagination run free. Don't hinder it with false insecurities or thoughts. Your mind is lying to you by presuming to tell you what you need. This exercise is your moment to tell your brain what you want.

2. Recurring Memory Theme

Choose a recurring theme that keeps harassing you, something that encapsulates a lot of memories with your ex. For example, I chose bars. My ex and I used to try out many different bars looking for happy hour specials and checking out breweries with friends. Replace it with something you know you like, such as a restaurant, wine tasting, or nightclub, or remove it entirely; to each their own.

In my case, I'm a man of simple tastes and couldn't care less about beer, so it's unlikely you'll bump into me at any breweries any time soon. On the other hand, if you are fond of new bars and happy hours, that's great. You don't have to change the activity, just change the place. Practice your writing skills and describe the scene as best as you can and introduce the hunk from point one into it.

3.   **New Activity/Hobby:**

What did you always want to do that your ex didn't want to? That shouldn't be too hard, I hope. But, you never know, and in that case, simply think of something new you want to do, perhaps something you've always thought about doing and haven't. This activity can be something big, such as skydiving (such as myself), or something small, like watching TV in your bedroom, or the movies you like, or jogging, or cooking, or even taking up biking daily with your hot *friend.*

When you finish, keep that page with you as a constant reminder. Read it every morning to reset your mind to a fresh start with your positive thoughts. We have to get the day started, not let the day start us.

Placebo Effect

"[S]he who says [s]he can, and [s]he who says [s]he can't, are both usually right."
Henry Ford

A placebo is simply a sugar pill. For example, nearly every prescription drug you've heard of on the market today was approved by the FDA after passing a placebo test.

Outcomes by the University of Colorado Boulder measured the neurological and behavioral effects the placebo effect had on a group of recently broken-hearted volunteers. Analogous to the Columbia University study researching the correlation between physical pain and emotional withdrawal, a group of forty participants were subject to like measures.

Participants were exposed to physical pain through electric shock, viewing pictures of their ex while thinking of the pain during their breakup, and looking at the image of a friend. After initial testing, the group was broken into two equal parts and given a nasal spray. Half of the participants were told it was a "powerful analgesic effective in reducing emotional pain," while the other group was told it was a saline solution.

When the participants were exposed to the testing process again, the placebo group not only reported feeling less pain but shockingly, their fMRI scans showed their brains also reacted differently, showing lower activity associated with pain.

That's the incredible power of the mind—it can be your best friend or worst enemy. The outcome lies in the decision, and the decision is ours. In other words, do what you think will make you feel better. Even though a lot of work has been put into designing and structuring these two breakup chapters for you as a guide to help you cope and understand your situation, there is no one single solution to the complexity of your mind through such experience.

Just as with our previous exercise of substituting memories by using the power of belief with our imagination, believing what you are doing will have a positive effect on your emotional status—science tells us it will! Go ahead and help yourself in whatever way you feel best works for you.

The Three-Week Rule

"You get a car! And you get another one, and you get a car!" I'm sure you're familiar with Oprah, but have you heard about her twenty-one-day strategy to form a habit? Well, the explanation is in the title—it says it takes twenty-

one straight days of doing something for the action to become a habit. And though others have their own theories, such as Tai Lopez, who states it takes sixty-seven days to form a new habit, the twenty-one-day rule has been more socially accepted by science and experts. Besides, Tai is a great entrepreneur, so taking his sixty-seven-day challenge may be better for when you are ready to turn into the next prominent business magnate.

I happen to concur with the former. In the same way, repeating an action for twenty-one consecutive days forms a habit, alienating yourself from specific actions and activities while sticking to your plan will make the heart-wrenching symptoms subside naturally. As if by a touch of magic, you'll find yourself waking up one morning to find your symptoms are simply—*gone.*

Listen, it may be tempting and easy to cave in and replace one body with another. You may miss being touched, feeling the warmth, and cuddling. But this choice will only hurt you in the future and can lead you down a vicious cycle that will have you questioning how you ever got there. Well, I can tell you the answer right now—it will be because you cheated yourself because you couldn't resist three weeks to stay away from the addicting pleasures of life.

Why is it so important? Because you'll be doing it out fear—fear of your cravings, fear of your pain, fear of your thoughts, masquerading insecurities with pleasure, dependence with fear of loneliness, without having the courage of looking inward. Courage is not being unafraid, but being afraid and deciding to face it. It takes courage to look inward, to be brutally honest with ourselves. A lie to the mirror is always more comfortable, and it often hides as a subtle "reason" pointing elsewhere: *The gym is too far…too expensive…it's not open long enough. I just wasn't happy. There was a lot of traffic. It just wasn't working out. Oh well, I tried. I did pretty well; it's all right.*

Happiness is a pursuit. Stop waiting for it because it won't come on its own. Secondly, happiness is a state of *own*. We must find happiness within first, and then the goals, people, and pets we choose to bring into our lives will emphasize that happiness, but will *never, ever* provide us happiness on their own. The tape over the leaky hole will need constant replacement. We must address the issue, be honest with ourselves, allow ourselves to heal, search until we find ourselves, and what's more, eventually fall in *love* with ourselves. Perhaps then, we can give the gift of ourselves to someone who's done the same.

"What" is the Question of

Love

Knowing my name is William J. _ _ _
You know what I study, but which of my studies,
reveals the rhyme that you're hoping to find?
The answer's quite formal, and of course it's of concerns of the
mind,
but the title doesn't follow the latter it rhymes with the former;
a name and a title making the secret untwine.

FOUR

What You Say You Want

and *HOW IT'S NOT QUITE SO*

I n 2000, Regency Enterprises produced a film starring Elizabeth Hurley and Brendan Fraser called *Bedazzled*. In this movie, a character, Alison, is a beautiful woman with dreams of meeting a great man. Despite attracting good-looking, smart, and successful men, she writes in her diary that she wants someone different.

"He's a man who's in touch with his emotions. He's a man who's not afraid to share his fears, his disappointments, and his tears."

As fate would have it, Elliot would come into her life. He sees her as the great person she is and respects her individuality and uniqueness. Furthermore, Elliot is in touch with his emotions—a bit too much, it would seem because his rival is Jerry, a douche who's the opposite of Elliot and only wants to get into Alison's pants.

Yet, Alison seems to find this desirable, as she later tells Elliot, "You're just too sensitive. I just want to be with a man who'll ignore me and take me for granted, who's only pretending to be interested in who I am and what I think so he can get into my pants."

Moral of this story? Jerry wins.

* * *

Women often tell me I'm lucky to be a man, but I've never quite understood why men "rule the world." I can't help but think we are powerless children under the thumb of a goddess who doesn't yet know she is one. It leaves me wondering why Wonder Woman ever needed a superpower? Perhaps men, society, and even yourself have submitted all your prowess into oblivion. It's not some fancy pitch or tool to seduce you with flattery; it's a genuine thought I've had for some time, and I'm expressing here for the first time.

Listening to the different voices of women from the world of dating, as well as those of close friends describing their personal experiences with men, I am confused why competent women, filled with prowess and beauty, would ever succumb to dating these type of men—men-children.

So, to the lady reading this page, when will you conquer the world? Why have you spared us from your sovereignty? You were born a prodigy and raised generations of men while fighting for equality when you were never equal to begin with—you are a superior design.

Perhaps after learning the secrets of love—the way men think, the five dimensions of compatibility, the five elements of love—along with essential relationship insight, and a dash of tough love, you'll fulfill what was meant for you to do since the beginning of your adult life—conquer. But first, we must learn, then realize, then we must accept, and after time we must apply. Once you have done that, you will inevitably conquer.

Lost & Confused?

If you've felt lost or confused at any point in your life when looking for that special someone, wondering *why*, you're not alone.

Yeah, right!

Searching, courting, dating, and jumping into a relationship can all be darn puzzling and hard. Every step of the process has different questions, schemes, and approaches. As you've read before, who said it had to be easy? Yet, no one ever said it had to be that hard, either.

When on the subject of love and relationships, I like to refer to a poem titled *Oh, The Places You'll Go!* by Dr. Seuss. The poem illustrates the uncountable decisions we have to make in our lives daily. I guess if we want to be technical, we *can* count the decisions we make in a day. The average human makes about 35,000 decisions a day. Multiply that times 365 days in a year, and then that times 79, the average life expectancy, and you get 1,009,225,000. Yes, that's over a billion, but who's counting? You don't get any extra lives, or a "Get Out of Jail Free" card, while always being a decision away from screwing things up. Piece of cake, right?

However, there is still nothing weird about you feeling a bit lost at times. Just think of the events that have to take place for two people to meet in the first place. Mathematically speaking, it is an extraordinary thing. Then, after making all the right decisions, being struck by lightning, and finding that lucky someone, you are only expected to have all the correct answers, make some more decisions, live, sleep, and wake together, be intimate, deal with the good, the bad, and the ugly, while having unique and individual thoughts, and stupid-strong emotions for each other. Oh, yes, and make it work for the rest of your life. Easy! Who would ever need a book for that?

It's vital to make a crystal-clear distinction between what we look for in *someone* and what we look for in a *relationship* as they are different things. We often get these concepts thoroughly confused. It is an ingredients-meets-the-recipe process, and as you know, the marinade on the chicken is as essential as knowing how to cook it. See it as what we look for in someone vs. relationship is to what we need vs. how to keep it.

What You Say You Want

Once, while having a conversation with a friend about life, love, the birds, and bees, she said to me, "You deserve better, Hugo. You really do. You're a very nice guy."

I pondered this for a minute, and let the silence take over the moment, then I answered, "Maybe that's the problem."

I remembered the harsh truth I once knew. You don't get the good guy because the good guy *doesn't* win. Unfortunately, the good guy finishes last. He never gets the girl and never has.

Yet after a bad breakup, treating people as I'd like to be treated, somewhere in between the journey of solitude, self-love, and dating, I decided to simply let go of all vanities and somehow become a good man. A man who was ready for who I hoped would be my last love. As I would soon find out, I would be in a failing relationship once again.

Then, where is the man of your dreams? Or does he even exist? He is precisely there—in your dreams. If you read some of the top articles after searching "What do women want in men?" or you scrolled through some online dating profiles, your social media feed, asked a few women (which I did), or answered the question yourself, you would describe the man of your dreams something like this:

"I don't have a 'man of my dreams,' per se, but I think he has some overall qualities. He is a good man, above all. He is a gentleman willing to show me chivalry isn't dead. Preferably good-looking or cute. A man who is in touch with his emotions and is not afraid to express his feelings and show his moments of vulnerability. A man who sees me as equal and is unafraid of an independent woman. A man who has his finances in order, at least as much as I do. Also, chemistry is crucial; it is natural both in sex and the relationship. And why not? If he has similar interests such as cooking, pets, and travel, that's a plus. Lastly, if he has a sense of humor and makes me laugh, he may have the key to my heart. Because laughter is the main ingredient in happiness and love."

"Human lies!"

– Koba from Dawn of the Planet of the Apes. 2014

Well, not entirely all lies, but to identify lies from truth, we'll have to dissect this further. Now, for the sake of simplicity, let's extract the elements from the description in the paragraph above:

- **A good man overall**   - **Sees partner as equal**

- **A gentleman**   - **Finances in order**

- **In touch with his emotions**   - **Chemistry**

Though you might find it hard to believe, the man described above does exist, and what's more, you've probably met him. Look around—he's probably somewhere in the pile of guys you consider good friends. On the other hand, if you haven't met him, don't worry. Keep your eyes open, ears sharp, and remember this paragraph because the signs will be there, and hopefully, after reading this book, he won't be joining the others in the pile.

If this is true, why is it that you're not with Prince Charming? Well, it's hard to say, isn't it? The truth is you're not quite sure. What you do know is that your inner voice may whisper in your ear, *he's so nice, but he's so boring,* or *he's such a great guy, but it just didn't work out,* or *we're just so different, and I need to be myself,* or *I don't know why, but I couldn't love him.*

When we fail to find detail in our reasoning, that's because the detail may not exist, and if it doesn't exist, the truth is, well, we are as blind as a bat. So, on the contrary, we justify ourselves to shut our inner voice up. We lie to ourselves, plain and simple. But why? Because we don't know enough about ourselves. So, what we think we know is not quite so.

Consequently, when belief, which can stem from an *erroneous assumption,* meets *reality,* you get → **Cognitive Dissonance**. Cognitive dissonance is a psychological explanation that means that when a situation presents something different from our thinking or beliefs, it creates an inner conflict, which leads to → excuses.

You'll hear about cognitive dissonance again in later chapters in much more detail. Anyhow, when we introduce excuses, we're left with two options: settle or stay single and dating, remaining completely freaking lost.

Though you may hate to admit it, thanks to this mayhem, the majority of ladies end up choosing one of three categories of men when they settle: a **fixer-upper**, a **safe bet**, or the **bad-guy**.

Warning:

Before you proceed, please refrain from reading this in the proximity of the opposite sex; I cannot be held responsible for any bodily injury.

The Fixer Upper

Women choose a fixer-upper if there ever was one. A fixer-upper is a man who compliments you in different areas but lacks some significant others. Maybe he's not the best looking, but he makes good money. Consequently, you get to work on his appearance and show him a bit about fashion and perhaps even pluck his eyebrows or take him to get a manicure while you're at it.

On the other hand, perhaps he is a good-looking stud, but not the sharpest tool in the shed (not much you can do on this one but feel empowered), or he might even be both smart and good-looking, but he has no ambition or drive. This fellow gives you the "opportunity" to motivate and guide him onto the right path. Women by nature have a nurturing instinct and will stay busy trying to close the gaps on the fixer-upper.

The Safe Bet

Let me introduce you to door two, the safe bet. This man is by no means a bad catch, but he's not the biggest fish to fry either. Sometimes, you might even hear the title in the description—"I like John. He makes me feel safe." Let's not get it twisted though, protection is a good quality in a man that women often seek. However, John is identified by certain characteristics.

John is not the most exciting character in the world, and while he treats you to your favorite pleasantries, let's just say the night is full of smiles and not so much laughter. John might also participate in adventurous play every from time to time, but he's not going to have you in bed with your eyes rolled back either. In other words, John is neither a party pooper nor the life of the party. He's more like the guy who shows up a bit late and lingers to help with the cleanup.

For the sake of stopping the bullying onslaught on poor John, he is, by the full definition of the word, a good man. One who is nowhere to be seen while you're in your twenties, *or is he?* Regardless, John probably comes into the picture in your mid-thirties and above.

Ladies mostly settle with John because they feel they're "running out of time." That unfortunate thing that gets us all—age. Every so often, I find myself facing moments of awkward silence when a lady in her late twenties says, "Ugh, I'm getting so old," followed by some more self-criticism: "I have gray hairs...I may never get married...Ugh, I don't want to think about it." Hence, why you don't ever ask a lady her age. Meanwhile, I'm about to turn thirty-two and feel fresh as a cucumber. But why should you and I feel any different?

Unfortunately, due to a bullying, misplaced social bias, women are commonly pressured into dating someone older than them and are also guilted into feeling like a failure if they haven't found love by a certain age. Based on the 2014 Current Population Survey, the average age difference (for a heterosexual couple) is 2.3 years, with the man older than the woman. In 64% of heterosexual couples, the man is older. In 23%, the woman is older, and in the remaining 13%, the partners are fewer than twelve months apart in age.

On the other hand, and much to women's credit, another common factor in settling is the poetically beautiful element of a woman's biology—creating life. Most men envision having kids at some point in their lives. But women are the embodiment of life, the mothers of the world, and in all that magical beauty, how could we ever possibly begin to comprehend what that yearning for having children feels like? Men could never understand this implicit pressure. So try to stop punishing the woman inside, and carrying the weight of social standards, and try to embrace the growth and gift of who you are today.

"Yesterday is history, tomorrow is a mystery, and today is a gift. That's why we call it the present."

— *Bill Keane*

The Bad Boy

Last and most certainly least, let me introduce you to the a*****e. Oh, never mind—you've already met him. If that word is not quite your cup of tea, allow me to translate—a douchebag, a d***head, a jerk, or otherwise more commonly known as a "bad boy"…at best. Women who find themselves with this character will be even more clueless as to why they are with him, aside from finding themselves saying *But I love him,* or *he's just so hot.*

Bringing such a personality into your life could prove rather unhealthy. This type of guy has been around the block once or twice and has developed a system from what he knows will work to get into your pants, and while all of us men primarily want sex, this Casanova wants *only* sex. In bed, he is dominant but lacks intimacy and may even be so forward as to voice personal requests. The bad boy is all about charming you, creating an atmosphere for the sake of giving you an experience you'll like but won't last.

He's usually in his mid-thirties, sometimes forties, he's most likely in decent shape and often has an apartment that looks very much like a bachelor pad (I would know). All aspects that you may find highly attractive, and while I'm not making a sweeping generalization of all men whose life fits into that description since they are good qualities, I am raising a word of caution for you to proceed *with* caution.

He also happens to take a decade to respond to your messages, and when he does, he responds vaguely. He rarely compliments you, and when he does, it's somehow sexualizing your physical appearance or praising specific worldly things such as your outfit, hair, nails, or style, among other things. Which are all things you may want to hear, and yet mysteriously nothing any heterosexual male has ever given a f*** about. Now, you might be thinking, *Not all men are like you, Hugo.* Touché, just don't say I didn't warn you.

Yet, something tells you he almost sounds too good to be true. It is a little, tiny inkling that makes you think, *hmm, he looks like he might dig himself more than me.* He does. You might even find yourself smiling from ear-to-ear this very moment. The most uncomplicated advice to give you is—*Run!*

It's all very resembling of Little Red Riding Hood, and if you need a quick refresher on the story, it's best summarized as: she was as clueless as a Polish detective! So, don't be Little Red Riding Hood. Chances are, that if he seems too good to be true, and all characteristics fit nearly perfectly, but he lacks intimacy, he *is* too good to be true (I wasn't always an angel).

If you are purely looking for some adult fun yourself, go ahead. But why kid yourself? Chances are, you might not want the sex as much as you want the validation that you are sexually desirable to replace the emotional rejection of feeling unloved.

"All right, so let me see if I have this right. If it isn't that we don't know, but that what we know is wrong because we fail to understand ourselves, and by doing so, we end up either settling or single and alone, endlessly dating, *how are we accomplishing this? How do you explain such a phenomenon?"*

Bravo, excellent question. Let me take you to front-row seating to one of the greatest clashes of life's existence.

Biology vs. Consciousness

Biology wins! I should have said, "spoiler alert." It's imperative to properly understand this concept to grasp what the entire chapter is trying to illustrate: how what you say you want; it's not quite so.

For the sake of simplicity, it means our human consciousness has gotten in the way of our biology's purpose. In our ignorance towards our goal, we assume and ultimately fail without knowing why. To explain this, we're going to have to embark on a journey through time, a journey of life. Strap in and enjoy the ride.

Life's Design

Our story begins millions upon millions of years ago, with a protagonist we will call Life. Life happens to be an engineer. This engineer's sole purpose is to create living organisms.

Thus, it embeds the first organisms with two basic, yet brilliant instructions in their complex code. First, everything must eventually die to feed new life, consequently creating an eternal cycle. Secondly, it programs all organisms with the intelligence and instinct of survival. This response would allow microorganisms to accomplish a few things. To continuously move life forward, it allowed the organisms to branch out into different types of organisms to create a variance for a magical symbiosis. Lastly, it would only allow the fittest to survive for life always to improve. Ergo, only the strong survive.

Like so, through the eons of time, the plan worked in perfect harmony, creating plants, bacteria, protoctists, fungi, animals, and viruses. Yet, among all these living things with the ability to reproduce, animals had a specific and interesting trait—sex. This simple, three-letter word has the complexity within the context of its name that could fill entire libraries. Once introduced, the battle of the ages commenced. On Mother Earth, the survival of the fittest was the new norm, and natural selection would become the study among thousands of scientists for generations since the most impactful publication on the subject known to humanity, *The Descent of Man, and Selection in Relation to Sex* by Charles Darwin.

If we dove into the nitty-gritty aspects of biology, we could spend the entire book and some change on the tiniest of topics. It is a subject so wonderfully complex that scientists can spend an entire lifetime studying one single insect. But I'll leave those extensive publications in the hands of far more qualified people. Yet, on the upper, thinner layer of the subject of biology, one sees certain common and dominant traits among the animal species—for example, the control and dominance of mate selection by females.

Throughout the animal kingdom, you'll notice the designed pattern of life in specifically choosing females as the symbol and embodiment of creation, holding the key to the door for which, on the other end, males will compete—often to the death—to be granted entrance. Telling us that if there is such thing as an engineer, males were *not* the chosen image or blueprint upon which life was erected. At least, from what evidence we can gather, biology points towards a prime suspect, the perfect creation on which all life was modeled—females.

That biological notion can be so fascinatingly captivating that when you narrow the scope on some of its animal species, it only gets amplified. In the world of insects, a male sacrifice is often warranted upon survival circumstances. The black widow spider, for example, sometimes kills—or, best said, *eats*—her mate after mating as a lurid ritual to gain his nutrients, giving the infamously famous spider its name. Another insect commonly known for its gruesome mating practices is the praying mantis. As if taken out of some macabre Stephen King novel, a praying mantis at times will also kill her partner during the mating practice for the same nourishment reasons by cutting off his head. Interestingly enough, males killing females is nearly nonexistent.

The same can be said about many of the vast arrays of different species throughout the insect world: the sagebrush cricket, different spiders, and the evolutionary survival of the Blue Moon butterfly. Yet, even when mating doesn't prove to be a deadly affair, you can see the female hierarchy pattern. Likewise, all working male ants punch in and out of work daily for their ant queen. Similarly, hardworking, fluffy little bees work for the queen bee, who doesn't even have to leave the beehive. Even some humans are catching on with similar fluffy, yellow online dating apps.

On a grander scale, in the world of amphibians, reptiles, and mammals, the female dominance trend need not change. If we take lions, for example, one would first be safe to assume that the power shift changes, since it is the male lion who has a pride of lionesses to mate with all to himself and spread his seed. And even though that is absolutely true, it is the male lion who will compete and fight to the death with another male who approaches his pride. The champion keeps the pride, and the loser becomes a stray feline shamed with a self-guilt so high that his testosterone levels drop, eventually thinning out his mane.

Yes, the lion loses hair upon losing the fight. Intense stuff, right? Not only that but if it's the newcomer male who is the victor, he will kill off the pride's cubs to ensure his seed will be the one passed on. Similar sexual competition among males can be seen among other species as well, such as gorillas and even reptiles whose female counterparts always choose the strongest, most dominant, and fittest of males.

But whether it is a male sacrifice, male workers for a queen, or male competition to establish dominance, somehow, the design, and purpose, seems always to be the same. Males spread their seed, and females have the power of choosing, holding the keys to natural selection, as life's HR department to find the fittest of males.

As we commonly say, life was good, and though biology's cycle of natural selection and survival of the fittest may seem cruel to some, biology doesn't care about our thoughts or feelings. Under such rules, living can be an arduous task, and biology may seem like an indiscriminate murderer—because it is, and only the strong survive. But before long, life would take on a new project and would employ a new design.

Sometime around a hundred-and-fifty to two-hundred thousand years ago, life by some unknown reason would branch out a new species of mammal. One that would eventually put biology's natural murdering streak to shame. What motivates this new creation? Perhaps it was done as an experiment. No one knows, but if not a failed experiment, most certainly a dangerous one. I guess, to some extent, the phenomenon still leaves some leg room to wonder if this living organism is an animal or a virus. This species would come to be known as Homo sapiens.

Enter Consciousness

The fact that we as a species acquired our level of consciousness is beyond fascinating by the simple fact that it had never happened before, and it has never happened since. To what purpose or why it happened, we do not know. There are many theories. Today over four thousand different belief systems or religions exist. All the world's sciences and all the greatest minds in history combined still have not found an answer to why we acquired consciousness. Even I'm writing a theory into a book trilogy, attempting to shed light on that question. Now call it God, call it life, call it science, the ether—that's beside the point. But make no mistake about it; we are *designed!*

What we do know is that upon the introduction of human consciousness, the cycle of life explained above would be permanently disrupted. How so? Well, consciousness jumped humans from being a weak animal among many, to the very top of the food chain. In its climb to the top, humans eradicated all other human species—Homo neanderthalensis, Homo heidelbergensis, and our own mother, Homo erectus, along with over 95% of all different animal life. Just ask the poor ghost of the dodo what his recollection of humans is.

This unprecedented level of violence by the introduction of humans and their "consciousness" was something for which life on planet Earth hadn't evolved enough to prepare. For the first time, a living species would go from surviving to complete domination. For the first time, a species would go from surviving to thriving such that it would duplicate its species' numbers yearly without an end in sight.

As a result of such dominance, power, and subduing of the biological order, the human race became complacent. By extracting some of the examples above from the animal kingdom and comparing the biological trends in the empirical data versus modern humans, it becomes quite clear the root of the problem of why humans fail at love is that we are gifted with too much power. So powerful, we get away with the bare minimum. Whereas every other animal and plant that has ever existed must give it their absolute 100% to compete and keep its place in the chain of life, and even when putting forth full effort, survival is not a certainty.

Humans, on the other hand, get by merely giving 20% (non-scientific, estimate to illustrate the point). For the first time in life, because of our dominance, the weak weren't destined to die but be dominated by the strong. This event would formally subdue biology's programming—survival of the fittest—to its knees.

At such toxic levels of complacency, you get a human species ever so confused about our purpose in life, depressed, feeling lost on an endlessly floating rock, without finding answers to our many questions, becoming imprisoned, and unable to escape our frustration; thus, our cry for equality among ourselves. Overall, the introduction of consciousness led to an awakening in which listening to biology's programming was not enough again. Therefore, consistently conforming to our natural selection and after wandering through an overwhelming and chaotic maze of consciousness, surrendering to the thought of feeling—unloved.

Humans are not a specific alien creation. We are a biological creation by whoever we want to believe, also in our own right. Yet, and most importantly, one out of thousands of species sharing this planet. An animal by mere definition, and a mammal at that. We are unique and beautiful in our own way and right, despite our many flaws, and still, the rest of the animal kingdom is magnificent as well.

But we fail in comparing everything to our intelligence and consciousness, and when we adopt that instrument of measure, well, nothing quite compares. Because if that's the case and we judge an elephant by its ability to climb a tree, there's a vast ocean of things and skills that other animals possess that we can't do either.

Despite all of our gifts, freedoms, and attributes, we are also governed by an explosive concoction of chemicals. Chemicals like kisspeptin, luteinizing hormone, follicle-stimulating hormone, estrogen, progesterone, gonadotropin, thyrotropin, among many more, contribute to such concoction. All these chemicals are created from an umbrella of primary elements like carbon, oxygen, hydrogen, and calcium. These elements create a biological program intended for us to think, question, learn, believe, imagine, socialize, reproduce, bond, find each other physically attracted, have sex, cry tears from emotions such as fear, hate, happiness, pain, and love.

Despite our consciousness's attempt to deter from our biological instinct, we are still a design obeying a set of rules, compelling us to compete, choose the most dominant, strongest, and fittest. Being conscious beings, we carry a lot of responsibility on our shoulders. All of which can be somewhat confusing to understand and dissect when we think we know something, and our biology calls for natural selection. Thus, responsible for making you walk away from that gentleman, who is in touch with his emotions. Consequently, decisions that lead you to a lonely path wondering how and why.

(Read from left to right and top to bottom)

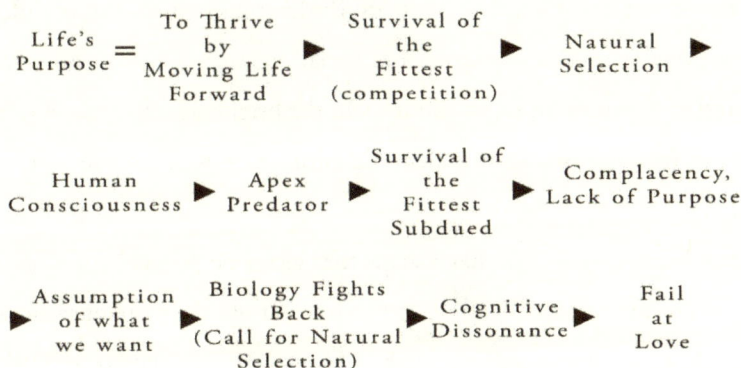

Life's Purpose = To Thrive by Moving Life Forward ▶ Survival of the Fittest (competition) ▶ Natural Selection ▶

Human Consciousness ▶ Apex Predator ▶ Survival of the Fittest Subdued ▶ Complacency, Lack of Purpose

▶ Assumption of what we want ▶ Biology Fights Back (Call for Natural Selection) ▶ Cognitive Dissonance ▶ Fail at Love

In Love or In Lust?

Upon learning the pattern of life, understanding our biology, and comparing it versus the rest of the animal kingdom, why would our purpose be any different? After all, we possess qualities that suggest we humans also have an overall objective that goes beyond our often narrowed notion.

Charles Darwin supported this claim to an extent in his publication *The Descent of Man, and Selection in Relation to Sex* when he argued aesthetic traits drive sexual selection preferences. Many of his contemporaries argued against this, and thus, this specific claim remained in obscurity for about a century.

The subject of human sexual selection is extensive and possesses arguments from different angles that provide a variance of different perspectives. An article published by Michael Lawrence Wilson Humans as a model species for sexual selection research states that upon searching the term "sexual selection" on the Web of Science for publications, you get 27,033 results, of which 98.5% of them were written after 1990. So, that should lend you some insight into the complexity of the subject of sexual selection.

As humans have advanced socially, the empirical data on the subject far exceeds that of any other living species. But, because humans have different cultural practices, belief systems, and social practices, drawing an absolute conclusion is simply an impossibility. On the bright side, though, it doesn't deter us from gathering prevalent denominators to illustrate our point further.

For example, the penis has a shape that serves no purpose for pleasuring our female counterparts. It's safe to assume that upon a woman seeing one, she might have a few questions, such as hmm, *who designed this? And why did they give it such shape?* I agree with you. It's when you find out its design is for the sole purpose of biological competition that it all makes sense. The penis is designed to drag out and extract the possible semen of other males. From there, the race to the egg begins. By the way, there are about 20 – 300 million sperm cells in a single millimeter of semen, so as you can imagine, it's quite the marathon.

On the other hand, women have a monthly visitor commonly referred to as a menstrual or ovarian cycle. As you know, you may notice increased libido during your ovulation. Though a woman's ovulation cycle is a complex process that involves eight primary hormones communicating or motivating one another to perform a given action, we'll direct our attention to the Luteinizing Hormone (LH) produced in the pituitary gland (along with many other hormones). Its secretion stimulates the release of an egg during ovulation two weeks into a woman's cycle. Studies suggest that women generally report increased libido three days before the surge of LH.

Furthermore, LH has several other functions, from stimulating progesterone to sustain a pregnancy to stimulating the production of Oestradiol. Therefore, the consensus on this marvelous ovulation process is that it might be responsible for your flirtatiousness toward your male counterparts that men love so much. Trust me, no one's complaining.

1.618—does that number look familiar to you? It's called the Golden Ratio. It's all over nature, all over our planet, and all over the universe, for that matter. It's also the simplest example I can show you to illustrate we're designed. Which bears the question, why were we designed? The question itself amplifies the excitement. The fact that you and I follow the same set of rules with mathematical precision is a fascinating concept. If you're willing to test it, try this out:

Simply divide your height by the length from the floor to your belly button.

Height in Centimeters

Belly Button Height in Centimeters (long measure)

=

Voila!

Consciousness Meets Awareness

Upon hearing an opposite argument, we can often make the mistake of assuming the preposition should replace the original point, making it obsolete. This notion couldn't be more wrong. The point is not to say that listening to our biology should override the introduction of our consciousness, but instead that we should find a balance between the two. As you'll learn in chapters ahead, two rights can coexist, and this happens to be exactly the case between our protagonists, Consciousness and Biology.

Before we proceed, the point is *not* implying that we should regress in our societal advances. It's also not saying that men should be free to roam and spread our seed as promiscuous macho playboys, nor that women should adopt the role of being caretakers because our biology suggests sexual and physical differences—of course not. Only that we should be conscientious about our needs to make better decisions.

Whether we'd like to admit it or not, from whichever angle you look at this puzzle, from a mathematical perspective, a biological one, or anatomical one, we are indeed a design under a set of rules, serving a purpose, and most certainly following a pattern. But just because we follow a set of rules doesn't mean we're not unique or beautiful, nor does it suggest we are no more than a bunch of zeroes and ones on a page. On the contrary, it should further confirm the conviction of our beauty, wonder, and magical existence in the cosmos.

If you'd like to know about your uniqueness, by looking briefly at the basic principles of cryptology[4], a simple, four-digit combination using the

---

[4] The art and science of making and **breaking** ciphers (a secret way of writing number combinations—algorithms.

numbers 0-9 has 10,000 possible combinations. Stay with me—just by switching to our 26-letter alphabet to form a simple three-letter combination without repetitions, that number of possibilities dramatically increases to 67,108,863. So imagine what the number of options is to create such a magnificent design as yourself with our DNA, which has 3.2 billion letters of coding.

Yet, if you are dying to know, a *Forbes* article written by Drew Smith, Ph.D. in Molecular, Cellular, and Developmental Biology tells us, "Four hundred twenty billion different variants are possible." Oh, that's in one generation, by the way. Yet, if you take the eight billion sperm cells generated in a sperm generation cycle multiplied by the average number of times we have sex, multiplied by the 20-300 million sperm cells per millimeter of ejaculation multiplied by the years of our fertile window multiplied by the variance in our ancestors times, *Wah-wah, wah-wah, wah-wah.* It soon starts to sound like Charlie Brown's teacher, right?

I have the number though, ready? That's $10^{3,480,000,000}$ possible combinations. The possibilities are in the trillions. Let's say the mathematical computation on a regular calculator reads *Not a number,* adding an incomparable value to your magical existence.

Yet, we still don't know why we follow such design. We only know what the design is (to an extent) and what it implies. From men having higher testosterone levels and women having higher estrogen levels, this pattern is presented in men as beards, lower voices, broad shoulders, the purpose of the penis, and overall masculine features.

The same can be said about women with their higher-pitched voices, breasts, broader hips, and overall feminine characteristics. As well as differences in height, size, and strength, and even the not-so-scientifically claimed and heavily criticized subject of pheromones. Though there's nothing substantial in data about this, hey, I love the natural scent of a woman—in fact, it drives me crazy, and I couldn't tell you why.

So, what could be more wrong than saying we know what we want, and what could be righter than saying it's not quite so? Yet, the purpose of gaining this knowledge is for consciousness to meet awareness. In all simplicity, it all serves to illustrate who we are from a scientific perspective, not who we should strictly be.

This chapter aims to aid our understanding to make better decisions by teaching you what our instinctual needs tell us about us, *not* that we should follow those instinctual needs without rhyme or reason. It serves to encourage you to engage your awareness, because what could be a bigger insult to our consciousness than choosing not to be conscious? So, if with great power comes great responsibility, let's embrace that power, but let's not forget to be responsible. It's merely a matter of choice because when ignorance ceases to exist, our excuses get flushed down with it. The decision is up to us.

Otherwise, what happens? We are doomed to repeat our own mistakes and engage in the worst kind of failure—the failure of self-fulfilling prophecy. Followed by a never-ending internal and external fight—what I say I want but doesn't fulfill me. What's worse, once we fail to look within, the battle between the sexes becomes an unstoppable force meeting an unmovable object.

Do you remember the solution? First, we need to learn, then understand, then we must accept, and with time, we must apply—then you'll inevitably—*conquer.*

FIVE

What Women Really Want

LAWS OF ATTRACTION

I
t is only after emptying the glass of perception and understanding the
love language within that we are free to rewrite the pages of interest
and dissect what we really want and need.

What we say we want is a product of what we feel and our environment
and not so much what the person inside is telling us.

No animal in the kingdom has ever been as social as modern humans.
This is due to the rise of consciousness and one specific, gigantic element
that only sapiens possess—imagination. *But don't other animals possess
imagination as well?"* Indeed, other mammals have a *form* of imagination.
Even groups of chimpanzees have been studied to have the ability to trade as
sophisticated and intelligent advances like trading food for sex. *What's
unique about human imagination then?*

You might recall the term nucleus accumbens, which is a part of the
brain responsible for our reward system—dopamine, from chapter two. The

nucleus accumbens forms a part of a bigger area of the brain called the limbic system, which is associated with memory and emotion processing. Human imagination differs from any other sort of imagination ever possessed by any other species in one thing, the power to imagine that which we have never seen, the future—*fiction.*

While a chimpanzee or other ape species can make complex trading systems with food, sex, and even tools, it is all as a result of information learned through different stimuli by their senses. Humans, on the other hand, can imagine planets and future societies and discover math from their advanced power of imagination. The introduction of this new power constituted the creation of fiction, which led to a more significant power, and perhaps a bigger problem—**collective imagination.**

In his book *Sapiens,* Yuval Noah Harari, Ph.D. in History from Oxford University, explains, "Fiction has enabled us not merely to imagine things, but to do so *collectively.* We can weave common myths such as the biblical creation story, the Dreamtime myths of Aboriginal Australians, and the nationalist myths of modern states…Sapiens can cooperate in extremely flexible ways with countless numbers of strangers."

The collective imagination is basically what its words imply—an imagined idea of fiction believed by a collaborative group of strangers. For example, religions, banks, companies, and governments. In most literal terms, a "bank" does not exist. You cannot ever meet it physically, nor can you meet a "government," or the base ideology of any religion. Their power is only gained because we believe it collectively.

In contrast, people like you and me are very real. You can physically meet us; we were born, and we will eventually pass away. Through millennia humans have created social norms and standards based on these collectively imagined entities. This fact puts an enormous amount of social pressure on us to behave, act accordingly, get a good job, be *fun*, get married, be an active member of society, be law-abiding citizens, and even *think* in an orderly fashion.

All of this consequently puts an enormous amount of pressure on the individual you to fit in. It's in our social nature to want to be accepted by our peers and be considered part of the social conceptual status. Because whether you became an engineer, or a firefighter, or a teacher, or chose to attend graduate school, or acquired your Ph.D., or have taken on the responsibilities of the family business, it is improbable you chose that path without the notion of social status.

In our love life, this dynamic does not change. Sure, you may very well want some cute kids you can name Autumn and Michael, and you're honest in saying you want someone with whom to laugh and drink merlot and zinfandel watching the sunset until you stain your teeth. Perhaps visit that city described as a moveable feast or eat oysters by the sea to the sound of waves crashing on the shore. But you, like me, and nearly everyone around us, most likely are not impervious to social pressures. Naturally, the stress we experience with everything else as part of our daily lives is not discriminatory of our love lives.

As we proceed through the chapter, you will read about a high-quality man, attraction, and chemistry, causing things to get a little confusing. Therefore, since that's the last thing we want, let's visualize something together first.

What do women really want?

To answer this question, we need to break it down into three further questions:

- What attracts you to someone else?
- What makes both of you compatible?
- What creates chemistry between the two of you?

Thus, **Laws of Attraction**, **Causing a Chemical Reaction**, and **Dimensions of Compatibility**.

Still with me? Great, keep following along. The chart below breaks down these components into categories. Take a look.

What do Women Really Want?

Laws of Attraction	Chemistry	Dimensions of Compatibility
Attractiveness	Conversational	Openness to Experience
Intelligence (IQ)	Personal Interests	Conscientiousness
Personality	Sexual	Extroversion
		Agreeableness
		Neuroticism

All right, easy-breezy so far, right? Let's focus on the Laws of Attraction first.

Laws of Attraction

Attractiveness simply refers to the biological factor that makes us physically attracted to someone else. But for the sake of not being overly systematic, I'll break down attractiveness into physical attraction and primary traits that attract you to that lucky someone in what we'll call *Traits of a High-Quality Man.*

Intelligence (IQ) is as straightforward as it sounds. The more equal the IQ level of your potential mate is to yours, the more compatible you'll be. Intelligence is difficult, though, because the only accurate measurement of intelligence that exists requires taking a test known as the Wechsler Adult Intelligence Scale (WAIS) or otherwise known only as an IQ test. But that might be a bit much for the first date, don't you think? Therefore I'll break intelligence down into the correlation with financial stability under the same section mentioned above, *Traits of a High-Quality Man.*

Personality is a thoroughly complex subject since it involves answering the overall question, *who are you?* Not to worry, though; experts have spent a great deal of time deciphering this into The Big Five Personality Traits. I'm going to share it all with you in our next chapter, "What Makes Us Compatible?"

Lastly, after you've read all about the Traits of a High-Quality Man, you can also expect to find a breakdown of what chemistry is between two people in our section, *Causing a Chemical Reaction* further down. Got it? If you have any questions, feel free to shoot me an email or a direct message through my Instagram page.

All right, let's get started.

Traits of a High-Quality Man

- Physically Attractive – An Alpha
- A Provider – Financial Stability
- A Gentleman
- Emotional IQ

Physically Attractive – An Alpha

Do ladies care about looks? Sure, to an extent, but women are usually not as drawn to physical attraction as men. Though women appreciate good looks, it's not really at the top of the priority list. It's more like the ingredient to a recipe that has a note saying you can replace it with something else or can exclude it altogether.

Have you ever seen that couple where you can't help but notice the apparent disparity in the aesthetic department? As a lady, upon seeing them, you might find yourself wondering what his personality is like before finding a smile on your face, meanwhile wanting to ask another question. Men, on the other hand, might be a bit blunter in their approach with a dash of obvious envy and straight out say, "I bet he has money."

Secondly, good-looking men are often seen by society as womanizers. The truth of the matter is that the handsome man is not as fortunate as you might think as he's been chastised by those same looks and labeled for decades as a Casanova, a Don Juan, or a complete Lothario. Though it might hold in some cases, you'd be surprised to know the desires and feelings of the handsome fella are no different than the next guy. Sure, good-looking men naturally get more attention, and statistically even have a 15% higher chance of getting a job after an interview. This trait can indeed contribute to higher self-esteem, but also meeting higher expectations. Yet it has little to do with being a douche or a womanizer. This is not to say women are not attracted to handsome fellas. It just means women and men are wired differently, and this happens to not be at the top of the priority list.

Nonetheless, biology, since the publication of Charles Darwin's *The Descent of Man and Selection in Relation to Sex* has made long strides for us to comprehend better what physical traits attract women. The best way to describe the physical attraction of women to men is by explaining that women tend to be attracted to a V-shaped male body, otherwise known in science as an upper-body-to-waist ratio. Studies illustrate women's trends of attraction by preferring physical attributes such as broad shoulders, bigger chests, muscular arms, and also a defined jawline to a flatter, lower percentage of fat in the abdomen/waist area. This arguably translates into a woman subconsciously deciphering this to mean this is a male who is best fit for reproduction since these are characteristics generally attributed to a healthy, fitter male. This indicates higher levels of testosterone, which means a higher chance of offspring and strength for protection during the eighteen most vulnerable months of a woman's life—pregnancy and the first nine months of an infant's life.

But why do these specific attributes translate into a better protector? Well, in our modern-day, it goes without saying that due to martial arts and the invention of weapons, among other things, this is not entirely true, and you'd be completely correct in saying so. But remember these are generalized explanations based on science to explain why you *might* be attracted to specific physical attributes and not *rules* set in stone that apply to all women.

That aside, these physical traits in men are biologically seen as more attractive since they'd be better suited for combat. Even in professional combat sports such as MMA and boxing, there are different weight classes to even out the advantage that heavier fighters might have. Broad shoulders, for example, mean that the individual has more power to drive a fist forward. Which as a matter of fact, human beings are among the few species (including kangaroos) that produce a tight fist for combat to use our entire arms as a club of some sort. In essence, a heavier male has more weight to put behind a punch.

A Provider – Financial Stability

HAVE YOU EVER SEEN A SAD PERSON ON A JET SKI?

If you asked a relationship coach or did a search on the internet for compatibility in relationships, you would get an ocean of different answers. In our subject of compatibility, one trait that affects the compatibility spectrum—intelligence—is generally accepted by psychologists. Like most things in science, nothing is a hundred percent certain; it's simply not how science operates. What we look for in science is a scientific consensus based on empirical data, never certainty.

Intelligence is not only a primary factor in compatibility, but science tells us that since women tend to date men of equal intelligence or above, the more intelligent a woman is, the less likely she is to find a suitable companion (sharpen up, fellas). Let's say a woman has an IQ of 130. Meaning, that based on her intelligence level, 95% of men would already be disqualified from dating her, which sucks for her because we don't choose our intelligence (not really, at least).

In all fairness, though, the lady in the example above is pretty exceptional if she scored that high on an IQ test standardized every couple of years to have an average of 100. Which would mean she's a tiny percentage of the population herself. Despite that being the average, according to an article by healthline.com, 68% of people fall within a score of 85 to 115. So I guess all you have to do now is administer an IQ test to your potential mates—I'm joking (please don't).

While the level of intelligence is a definite measure of compatibility, and while you can certainly take a test if you're curious to see where you measure up, it is rather unlikely—not to mention awkward—if you were to do so for the purposes outlined in this section. Despite how convenient you may find having your personal filtering system, there's no reason for the long face. As mentioned previously, data points towards a primary suspect that correlates (for the most part) to intelligence levels—financial stability.

Whether we've been socially programmed or biologically programmed, we simply don't know. Perhaps both. But for millennia, since the times of hunter-gatherers, the male figure has been associated with providing food, protection, and shelter.

Besides, it's not even necessary to dissect the subject so profoundly. Why shouldn't a woman seek economic stability in her male counterpart if she simply can? Some time ago, when chatting it up with a good female friend of mine, she made an insightful point. She said, "Sometimes a girl gets used to things and compliments that become 'normal.' Like, dinners, really nice dates, and pampering. So, if a new guy comes along and doesn't do those things, she cannot help but draw comparisons."

It's important to stress that this doesn't mean women can't make their own money, just what data shows women prefer and gravitate towards. A publication by Miriam Gensowski from the University of Copenhagen and IZA tells us that *"IQ affect the levels of earnings, especially in the prime working years. They also affect educational sorting and thus command an indirect effect on lifetime earnings."* In the same report that was conducted by analyzing data from the Terman Study, Gensowski[5] concluded that women with doctorate degrees have higher rates of return than men.

---

[5] The American psychologist Lewis Terman initiated this study with boys and girls born around 1910.

It is the longest prospective cohort study in existence (Friedman et al., 1995a), and is described in Terman (1925, 1926, 1930, 1947, 1959).

Emotional IQ

BOYS DON'T CRY

But real men *do*. If I told you a secret, would you keep it? I certainly hope not. Because we've been keeping this one for far too long, and it would defeat our purpose of changing the perception of the status quo.

Studies suggest women and men have the same range of emotions. A survey from psychologist Kateri McRae of Stanford University, "Gender Differences in Emotion Regulation: An fMRI Study of Cognitive Reappraisal," states, "In fact, the belief that women are more emotional than men has been labeled a 'master stereotype' (Shields, 2003)." In other words, studies indeed confirm the myth of women being more emotional than men is precisely that—a myth. What women and men differ on is our reaction to those emotions. So why is it that our society teaches men to suppress their emotions?

Women often say they want a man who is in touch with his emotions. Though this is a possibility in advanced relationships, it couldn't be falser in new relationships or the dating stages of this ingenious world of love.

Men *are* emotional, but we have been socially or classically conditioned to suppress our emotions by society. I'm afraid a man saying otherwise would be saying so purely out of insecurity. Deep inside, most men often feel a drowned longing to express what they feel. Overall, there are three major factors contributing to men keeping their emotions to themselves: society, male hypocrisy, and women they date.

Just as with women and age, love and relationships, and the rest of the societal pressures previously discussed, men face an invisible enemy when expressing emotions. From coping with the backlash of a breakup, to dealing with financial stress, to showing something as normal as love, a man is expected to simply "deal with it" or "suck it up."

Recently when I was having a conversation about the subject with a close lady friend of mine, she brought up an excellent point. She said, "If I, for example, were to be in a public bathroom and I started crying, all the girls in there would approach me to console me or ask me what happened, despite what the reason might be. And I'm not sure that would be the same case with men."

I laughed in agreement. I wish I could say otherwise, but the harsh reality is, we can cry all we want to about how unfair it might be that our society and women drown the voice of feeling, but the truth is we are dealing with a level of hypocrisy among men as well. Given the example above, the thought of a man consoling another man who's breaking down in a public bathroom seemed cruelly comical. Sad to say, but he may even be prone to hearing hazing remarks. In case it was a top government secret, let me whistle blow that men *also* cry. Yet, to further highlight this hypocrisy, I can't recall ever crying in front of another man. So what does that tell us about us?

Not so fast, Ms. Simone de Beauvoir. Women happen to be the third accomplice in this crime. In the world of dating, any man knows that one of the most important rules of the game is not to ever, under any circumstance, say something sweet or touching, as this will be a free, all-inclusive, one-way ride to ghost town. Compliment her? Ghost town. Respond promptly? Immediately to ghost town. Express you like her, or how much you like her? Forget about it, no questions asked, straight to ghost town, might get a speeding ticket while you're at it. Tell her you're not available, cancel on her, talk to other women, and respond the next day? Head over heels for him.

Wait, what?

Yes. That's because something in your brain attracted you to a man who has options, and while I'm not condoning such polarity of attraction, I'm merely outlining the opposite man who gives you undivided attention gets deciphered in your mind as he must not be the fittest of males, hence not the best option.

The truth is you don't even have to hear it from me. It doesn't take more than a trip down memory lane to laugh at some of your own mishaps. On the other hand, I'm sure it won't be too hard for you to think of a friend or two about whose love life you've thought, Gosh, I genuinely love her but, um, *'Oh no, baby, what is you doin'?*

Don't worry, you're not alone. I've thought the same thing. It shouldn't take a world-renowned physicist to tell you about the algorithms of humanity to understand that sometimes all it takes is a little bit of honesty with a dash of tough love and a sprinkle of logic. It's not a men vs. women thing. On the contrary, that's exactly what we're trying to avoid, if not solve. Adopting that mentality will only lead us down the path of an unstoppable force meets an unmovable object, and we don't want that.

Furthermore, we're humans. It is in our nature to think and overthink more than it's necessary, yet ironically, few times do we have the courage to search within ourselves and see that maybe it's not the world. Perhaps it's us. And though I'm sure you might have had your fair share of encounters with a-holes who played on your emotions because they had other obvious intentions. But blaming a group as a whole is the first step to prejudice.

We've established in this book we generalize for the sake of showing perspective, to learn about ourselves, but in no way to crucify everyone for the same sin. Having said that, if it's true we are a product of our environment, if it's further correct we're not impervious to social pressures, and we empathize with a reflection of ourselves—then perhaps what you mean by saying you want a man in touch with his emotions is, that you want a man with an emotional IQ—a man who will primarily listen to *your* feelings. And sure, it wouldn't hurt if you remember from time to time that he's human, too, and has a boiling pot of emotions as well.

Be that as it may, it's no secret men are not what you would call professionals at listening to emotions. With that in mind, cut him some slack. The same study mentioned above also noted that despite having the same range of emotions, men do react to them differently, whether it's because men are a product of their environments or something underlying that. Due to a life of practicing the suppression of such emotions, perhaps he has grown to ignore them a little better than you. Meanwhile, you probably just want him to express a word of affection every now and then to show he cares.

Thus, we have to understand and accept each other a bit better to meet in the middle.

A Gentleman

RESUSCITATING CHIVALRY

The "chivalry is dead" comments are prevalent as of late, and it's hard to object. Sadly, women couldn't be more correct on this one. Chivalry is most definitely dead, and it has been buried in the graveyard for years. However, men and women's fingerprints are on the murder weapon and with the compelling evidence, might as well skip the trial and plead guilty.

Men played a role in this murder with their machismo and rigid, ego-driven objectivism against the first two waves of the feminist movements, failing to recognize that women had a point. Naturally, the following generations of men sucked at what their predecessors mastered: chivalry, coquetry, courting. As women evolved, men have not so quick to catch up, and upon being faced with the more outspoken, rightfully demanding, and independent woman, men, are consequently intimidated. This sequence of events translates into a generation of men oblivious to the needs of women.

In this day and age, these men are known for their brute-force attack and "may the one who shows off more feathers win" strategy, and the let-me-flatter-you-with-a-hundred-compliments-a-minute competition began.

With time, women have become more and more immune to these gestures. Today, all it takes for a lady to get celebrity-level attention from men is being beautiful and flaunting her physical attributes. Thus, telling a beautiful woman she's beautiful is as flattering as telling her she's a woman.

As a result, eventually, men have slowly but surely adapted to this modern, more independent woman who is unfazed by flattery while simultaneously neutralizing their lame, unevolved comrades. Hence why, flattery, words of affection, prompt responses, and, sadly, overall honesty will forever be a thing of the nostalgic past.

But let's not forget men's loyal accomplice in this murder. Modern women and their pseudo feminism buried chivalry as well. I know that might have caused some irritation, so feel free to take a deep breath and try to proceed with an open mind. When feminism began at the end of the nineteenth century and beginning of the twentieth, it had a voice. Women questioned why it was that two members of the human race were not treated equally—and more importantly, why was it that the rule of law applied differently to both sexes.

According to Martha Rampton, a professor of history and director of the Center for Gender Equity at Pacific University, feminism can best be categorized in four waves. The first wave being the one described above, which involved the argument for voting rights and having a voice among men at the beginning of the century.

A second feminist wave came in the 1960s, known as the Women's Liberation Movement. The first wave of feminism argued for legal equality as human beings—explicitly voting rights. The second wave of feminism in the women's rights movement focused on elements of experience from a woman's perspective, such as work equality, having a voice in politics, their homes, their right to their sexuality, and family roles. These two waves can be held responsible for the most tremendous advancement in women's history, and women should be darn proud of it.

Then with time, something happened. Our society ended up in a waiting place where things were working, and men and women were getting along. But then came the third and fourth waves of feminism, otherwise known as modern feminism. While we, men and women, can sit here and point the finger at each other all we want to, endlessly arguing about whether it was the chicken or the egg first, it is beyond the point and solves nothing. In fact, it is that "who's right vs. wrong" competition that got us in our present love predicament in the first place.

What I can tell you is that somewhere along the line, we humans lost our voice of reason, and what was once a fight against the unfair treatment of women by men became a fight of power against men as a whole. Instead of fighting for understanding, compromise, and working towards a solution, our society became focused on winning by any means necessary. Ultimately, the word "equality" has lost its meaning, stubbornly clinging to a reality that was no more than a game of chicken between the sexes. All while missing the fundamental element, both women and men are looking for—Not equality, but fairness—a level playing field.

Today, naturally, this cry for equality is so cluttered and blurred that both men and women look like deer caught in the headlights, lost somewhere in the massive walls of a physical world in a Dr. Seuss poem.

Overall, both ladies and gents have been utterly and negligently responsible for the murder of chivalry. So, what could be more valid than chivalry is dead, and what could be falser than saying we want it?

The frustration is understandable from both perspectives. But despite feeling justified in our frustration, we have to use our conscious voice of reason while keeping in mind that two wrongs can coexist and that arguing solves nothing. But what's more, communication is about listening to each other and not about talking over each other.

For example, men expect to do certain things for women. Let's say it's a beautiful Sunday by the pool, and we're firing up the grill, barbecuing, and drinking some good ol' IPAs—*the works*. You pick up the grill, proceed to clean it up, set up the charcoal, and throw some steaks on it while you're at it. Yet, instead of a thank-you, what you encounter is a wounded ego in a grown man telling you he's got it as if you took away his favorite toy.

On the other hand, when a man decides to dust off his chivalrous side by offering to carry some groceries for you to share the load, what he encounters is a defensive snarl resembling something along the lines of, "I don't need your stupid muscles." So, the man is as wrong for expecting you can't handle some charcoal and grill, as you are wrong for being upset when he offers to carry some bags for you. The best solution is *to perform actions without a predetermined motive or notion.* How? By understanding the competence of women while not undermining the intentions of men.

So if a man is reading this, if she fires up the grill or takes up some groceries on her own, it's as simple as *she's an independent woman!* Don't overthink it. She's not taking from your manhood. Relax. Hopefully, there's much more to determine your manhood.

The same goes for the ladies. If a man offers to walk you to your car at night or carry a heavy bag for you, he's doing so because he cares about your wellbeing and is being a gentleman. Never did a quality man think, "Oh, this poor weakling and defenseless child must require my powers." Of course not. Don't overthink it either.

Putting our differences aside can be part of our solution. Listen, you've already read a whole chapter telling you about the many implications of biology, so there's no need for a comprehensive course on the topic. But denying our differences is nonsensical. The problem is we've blurred the meaning and intention of the words "difference" and "equality." Simply hearing the words now has a ring of implied superiority vs. inferiority—which from the definition of the words themselves, is *wrong.*

The entire purpose and beauty of us being different is for us to match like perfect puzzle pieces. *Not* see those differences as individual qualities as a reason to compare and compete. Make no mistake about it—if we genuinely want to invest in resuscitating chivalry, let's not forget that chivalry was born from our recognition of those implied differences in the first place.

In general, men are indeed physically stronger than women by the simple fact we produce more testosterone, which aids our muscular structure (not that you can see the frame in many cases). You are more symmetrical and aesthetically better-looking; you are more beautiful than men. Yet, who is more intelligent? None! So, what's the big deal? I am naturally attracted to your physique. You are more attracted to my strength, and we're both equally as potentially intelligent. I focus on your buttocks and your breasts. You focus on my broad shoulders and my deep voice (because I can't grow a beard). I, as I'm sure you can say the same, have no freaking idea why none whatsoever. The same applies if you're gay, you don't know. We don't choose or have an explanation for our sexuality; we only know that we do. I just know that such geometrical aspects of a woman's physique make me want to pursue you.

To shed some better light on our topic here, let's play a game of You & I Say vs. Our Brain Says. I look for the alpha female, and you look for the alpha male. Easy enough so far.

I say, "I'm attracted to the most femininely, physically attractive female with the best aesthetic features." In contrast, my brain says, "A good body must mean a healthy host to nourish a fetus for reproduction and good features that can be passed onto a healthy baby."

On the other hand, you say, "I'm attracted to his size (no pun intended), his strength, and deep voice." Your brain says, "Size and strength must mean a dominant protector for a higher chance of survival. A deep voice must mean a suitable level of testosterone for reproduction."

It's important to understand our biology has no idea what century it's in, nor does it care. Just like a chicken's biology (who's been around for a few more million years than humans) has no idea what century it is in, and that it has been invaded by humans. Otherwise, if it could speak, I'm sure it would say, *Run!*

Furthermore, as you might have noticed repeatedly reading, two wrongs can coexist, just as two rights can coexist. The fact that I like your feminine physique and focus on specific parts of your body does not negate the other 90% of your attributes. It simply means I like one of the *many* qualities that make you, you. Recognizing your attraction towards a man's strength, dominance, and all of the attributes those words encapsulate does not make you inferior or dull. It makes you human. It makes you a heterosexual female.

Once we dare to accept our different natures to a fair extent and embrace those very differences between us, perhaps we can look for our piece of the puzzle that matches us.

Otherwise, you're free to keep arguing that you are a better piece in the puzzle and effectively be left out as the missing piece.

Causing a Chemical Reaction

Chemistry, by definition, is the composition of elements that make up something else. In the formula of love, chemistry is essential when establishing compatibility. Ideal mates should look to have chemistry through three measures: *sex, personal interests,* and *conversation.*

FINDING CHEMISTRY and BUILDING RAPPORT

Looking for chemistry in a partner is not only vital but necessary. You and I may be ideal for each other on paper, but that doesn't mean we'll get along. Chemistry is more about the connection between A and B than it is about the similarities between A and B. In science, hardly can a chemical ever react if mixed with more of the same substance. See it as two chemicals that, when combined, elicit a reaction to transform into something else. As Carl Jung said:

> *"The meeting of two personalities is like the contact of two chemical substances: if there is any reaction, both are transformed."*

In my past relationship, one of my mistakes was allowing myself to put more emphasis on what she validated in me, instead of finding a connection between us. Thus, to find that perfectly compatible, matching piece to our puzzle, we must first establish *chemistry.* But it's also important to understand that while we find chemistry with another individual, it must be proactively sought and worked to build rapport. Because if you don't practice good rapport building habits, keeping a relationship down the road may prove challenging—which is not what we seek.

Sex is the door to emotional connectivity, bonding, and love. There's no better judge of your sexual preferences than yourself. Besides having an entire chapter on the subject as it's also one of our elements of love, showing genuine care for your pleasure is paramount. A man uninterested in your satisfaction is missing a couple of the factors that a long-term partner should possess. Sex is truly an art. There are few rules to it, and it should be based more on feeling and less on thinking.

Finding *common interests* will be part of your conversation as you're getting to know each other. While you may well know what your top personal interests are, jotting down three to five main interests may very well serve as a reminder to see if your date checks them off. Otherwise, don't overthink it. Remember, there are no rules to the world of love, only guidelines. We follow predictable trends to make better and guided decisions, but not one single person follows a specific algorithm by which to abide—at least not at our date and age.

Top Interests:

_____   _____   _____

_____   _____

Conversation is an essential factor in all stages of a relationship. It's the main conduit for establishing excellent communication and rapport. In the beginning stages of dating, the most natural way to distinguish good chemistry in this department is seeing if you two have naturally good, flowing conversation. There are solid habits you should observe in your prospective candidates that any good male communicator should possess. In business sales, for example, they have the psychology of good conversation down to an art and science. If you've ever been in sales, you may very well know what I'm talking about. Otherwise, here's your ultimate checklist to see if your partner is a good communicator:

1. *His greeting.* It should be warm with a genuine smile. A proper greeting will also form a complete sentence containing more than three words and will usually be followed by a question. Because if he's not trying or is clueless at the beginning, imagine when you're in a relationship. For example:

    "Hello, I'm Hugo, it's a pleasure meeting you. How's your weekend been?"

2. *He uses your name.* In sales, there's something called the "rule of three," which means that when we use a client's name three times throughout the conversation, it personalizes the interaction and, as a result, increases their likelihood of buying. Using your name doesn't mean he has to repeat it like a parrot, but when used sporadically, it is a good indicator of attention to detail and conversation acuity. Besides, he's putting effort into the interaction. So, give him some brownie points.

3. *He asks you questions.* There are two keys to this. First, a good communicator and conversationalist shouldn't only ask questions, but ask open-ended questions. An open-ended question is any question that requires more than a yes or no answer. Such as, *what did you do this weekend?* Vs. *Did you have a good weekend?*

4. *He listens.* A good conversationalist should *listen* more than he talks. Be on the lookout for two things primarily: *attention span* and *attention to detail.* To gauge his attention span, watch for his interruptions and that the common "yeah, yeah" is not too constant. A good listener has good attention to detail and should be able to repeat a whole sentence back to you. Repeating back to you, the last sentence you said, or the last words of a sentence, doesn't count. This dilemma is called the **serial position effect** in psychology—where someone is most likely to recall the first and last words in a sentence and forget the ones in the middle. Specifically, the tendency to recall the opening words is called the **primary effect**, whereas remembering the last words is called the **recency effect**. Good try, slick!

5. *He gives you undivided attention.* It's vital for you to notice your potential future partner gives you his full attention and is not so easily distracted. Wandering around, checking his phone, and even having his phone face down on the table are not great signs. Besides, placing the phone face down on the table hints to being protective over personal affairs (other dates), which is not the most respectful gesture. A cue to look for in a man with good communication skills is good eye contact. It shows you have his undivided attention and is an excellent display of confidence.

6. *He makes you laugh.* Two important things have been found in scientific studies about laughter. Aside from you knowing that you enjoy laughing, learning the reason why you laugh and what to expect from your potential partner may make all the difference in the world when gauging the qualities of that potential mate.

In a study by Jeffrey Hall, a professor of communication studies found by pairing 51 single heterosexual pairs that the more a man attempted to be humorous, the more likely his female counterpart was expected to be romantically attracted to him. Interestingly enough, the same was not true the other way around.

However, keep in mind a good conversation is a two-way street, and good habits should be reciprocal even if he's the initiator. A lot of women fall into bad practices thinking a man is instead an entertainer, and few things could be a bigger turnoff for a man. You are the queen, the goddess; no one denies that, but you want a man who knows his worth too. Falling into a game expecting the man to entertain you is silly, and you could be missing out on a quality match. If you don't share this thought, I'm afraid there may be more introspection to be done.

SIX

What Makes Us Compatible?

THE FIVE DIMENSIONS OF COMPATIBILITY

A s said before, in the complex world of dating, it's easy for us to make the mistake of missing the signs of compatibility—or lack thereof. But perhaps it's not as straightforward as making a mistake, but rather that despite knowing what we want, we simply don't understand what makes us compatible. So what makes you and I compatible—or not?

The good news is there is an OCEAN of information at our fingertips—**Openness to Experience, Conscientiousness, Extraversion, Agreeableness** and **Neuroticism**—otherwise known as the Big Five personality dimensions (or traits.)

The Big Five is a personality test based on five fundamental traits, the basic principles of which go back as far as Hippocrates of ancient Greece. But what is the Big Five, and how does it apply to you and your potential significant other? Well, for one, as in all things in life, knowing something makes us assess situations better and make better decisions. In the matter of relationships, the Big Five assessment is the most scientifically prestigious due to its accuracy. Well, again, does that mean we have to take a test? Not necessarily, but if you want to—for one—it's not as intrusive as an IQ test. Which, for valid and obvious reasons, may make people self-conscious about the cognitive abilities they're stuck with for life.

On the other hand, a Big Five assessment answers a more fundamental question—*who are you?* And what could be more important than your personality when trying to find its counterpoint? But no, it doesn't mean you have to take a test, but if you want to learn more about yourself, make sure to visit my page HugoBradford.com to get your results and see how you stack up against me. For now, learning about the five fundamental dimensions of personality should prove quite beneficial. Therefore, let's answer the following questions for a clearer picture: *What is it?* And *Why should I trust it?*

A Brief History of the Big Five

Let's focus on the latter first. Whenever I learn something new, my first response is to inquire more about it to test its validity. As I previously mentioned, the development of the Big Five personality dimensions goes back to ancient Greece to the times of Hippocrates, Plato, and his pupil Aristotle. They all came up with theories that broke down personality traits into four categories. At the time, the association of personality and brain was not commonly accepted. That came much later with the famous incident of Phineas Gage and the rise of a pseudo-science known as phrenology.

Phineas Cage was a construction worker who was victim to an explosion at work in which a rod perforated his left cheek and head. Cage survived despite the rod having destroyed a significant part of his brain's left frontal lobe, which altered his personality permanently. Yet, thanks to his unfortunate accident, the first link between the brain and personality was made, garnering worldwide attention.

This incident led to theories and studies from some of the most recognized names in psychology. Arguments such as Sigmund Freud's the id, the ego, and the superego, to the great mind of Carl Jung (Freud's collaborator/colleague) and his introduction of Introverts and Extroverts, followed by introducing four more essential psychological functions. Later, Hans Eysenck hypothesized there are only two determining personality traits—extroversion and neuroticism. Lastly, one of psychology's most renowned researchers, Lewis Goldberg, sculpted Raymond Cattell's sixteen "fundamental factors" into the five we know today. This model was later confirmed by two other prominent psychology researchers, Robert McCrae and Paul Costa. This widely accepted personality model is still the most convincing in our present time.

What this brief history should tell you is that the validity of the Big Five personality model you're reading about was a consequence of hundreds of years from some of the world's most prominent minds in psychology and based on data rather than theory. But what is the Big Five?

The Big Five

As we've discussed, the Big Five fundamental dimensions are Openness to Experience, Conscientiousness, Extraversion (also spelled "extroversion"), Agreeableness, and Neuroticism. What must be understood, though, is that these fundamental traits are not obsolete answers as they don't operate in a yes-or-no, close-ended fashion. Instead, they are measurements within a continuum of an overall polar spectrum.

For example: with extraversion, it's not as simple as being extroverted or introverted, but where you rank in the spectrum between the two. It doesn't mean you're one of the two, as no one can be purely introverted by the mere fact that we're human beings and social by nature. Otherwise, how could human consciousness ever have formed in the first place?

Also, the five fundamental dimensions of compatibility are composed of a forty-four-item inventory designed by Goldberg (1993). The Big Five Inventory looks like this:

Big Five Dimensions	Facet (and correlated trait adjective)
Openness vs. closedness to experience	- Ideas (curious) - Fantasy (imaginative) - Aesthetics (artistic) - Actions (broad interests) - Feelings (excitable) - Values (unconventional)
Conscientiousness vs. lack of direction	- Competence (efficient) - Order (organized) - Dutifulness (not careless) - Achievement striving (thorough) - Self-discipline (not lazy) - Deliberation (not impulsive)
Extraversion vs. introversion	- Gregariousness (sociable) - Assertiveness (forceful) - Activity (energetic) - Excitement-seeking (adventurous) - Positive emotions (enthusiastic) - Warmth (outgoing)

Agreeableness vs. antagonism	- Trust (forgiving) - Straightforwardness (not demanding) - Altruism (warm) - Compliance (not stubborn) - Modesty (not show-off) - Tender-mindedness (sympathetic)
Neuroticism vs. emotional stability	- Anxiety (tense) - Angry hostility (irritable) - Depression (not contented) - Self-consciousness (shy) - Impulsiveness (moody) - Vulnerability (not self-confident)

The Big Five Factors are (chart recreated from John & Srivastava, 1999)

Before we proceed with what the Big Five dimensions are, for the sake of clarity, keep in mind three important ideas:

1. Empirical data suggests a genetic factor meaning that roughly 66% of our personality traits are genetically inherited while our environment shapes the rest.

2. Empirical data also suggests gender is prone to a 60% variance factor, which means if you randomly selected a male-female pair, roughly 60% of tests would show that one naturally tests higher in that dimension than the other. The reason I mention this is so you're not alarmed when you notice this contrast with your partner; it is normal and expected.

There's also a majestic possible explanation to the variance, which I'll explain as we move along those specific points.

3. When evaluating compatibility between you and your potential partner, keep in mind these traits do not encompass a person's personality exclusively as personality, from what you can see, is a very complex subject. It is to give you a better understanding of who you and your partner are and measure your compatibility, not encapsulate you or them.

Likewise, a trait's rank compares to its polar opposite (ex., extraversion vs. introversion). It can also correlate with compensating or detriment one of the other four dimensions of compatibility.

When you find your significant other, it may be fun to take a test to gain a better understanding of your personalities. But as you go on your first dates, suggesting for them to take a test might not be the best ice breaker one has heard. But not to worry, to avoid this dissonance, what you can do is take the test yourself, and as long as you keep the secret between you and me, I'll give you questions you can ask to assess your candidates as you're having a conversation and building rapport. So *shh.*

Openness to Experience

This dimension is the closest association to IQ. In fact, it's often called openness to *imagination* or *intellect.*

These are your Neos and Trinitys (*Matrix*, 1999) of the world. Someone who ranks high in openness is likely to gravitate towards the arts, have a natural curiosity for learning new things, and possess unconventional ideas that challenge the status quo. It's more probable for them to become leaders as they tend to be unafraid of change. They are prone to being intelligent and are not great with established order and routine.

Openness is also the one trait least associated with any of the other four. It seems to be similar to processing ability vs. awareness. Where the ability to process tasks gives the impression of intelligence, it has little to do with awareness of the world around us and actual intelligence.

Questions for your potential mate:

- *What do you like to read?*

- *What do you like to think about when you have time to yourself?*

- *Do you care about art?* Or *What is your favorite art form?*

These are all simple questions that will give you some insight into the openness of your date. If they like to read fantasy or informative self-help books, such as the one in your hands, or if they like to imagine abstract scenes; and enjoy art, this person likely ranks above the median in openness to awareness. Compatibility depends on the comparison to where you rank, of course.

Conscientiousness

Here you have the individuals who have our economy growing and prosperous. The highly conscientious possess good organizational skills, self-discipline, and value authority. The conscientious individuals are competent in school, have compelling job performances, and the trait correlates with achievement.

On the other side of the spectrum, people who rank high in conscientiousness can also get in the way of their success if they allow themselves to lose focus and track of time by being extreme perfectionists.

Other things to keep a close eye on are that ranking high on conscientiousness has a negative correlation on depression, substance abuse, conformity, and seeking out security (Roccas, Sagiv, Schwartz, & Knafo, 2002).

A negative correlation to anything simply implies the relationship between two variables, in which as one increases, the other one decreases and vice versa. For example, if someone ranks higher on conscientiousness, they are likely to score high on agreeableness.

Questions for your potential mate:

- *At what time do you usually wake up?*

Someone who ranks high in this trait will probably give you a set time they wake up daily since they have an established routine. On the other hand, if your date says something like, "I try to..." or "Usually..." then you're safe to assume the opposite.

- *Do you like to plan when you travel or are you more "on the go"?*

While your date tells you all about how they loved visiting the Great Wall of China, you can gauge how orderly they are, based on how they plan for their trips. For example, I have great admiration for people who can backpack or roam and explore when they travel, since I like to plan and have a good idea of *what* to visit.

Extroversion

Extroversion vs. introversion translates into a more fundamental question—*from where do you gain your energy?* Someone who ranks high in this field puts a high value on friends, likes to start conversations, and values exciting activities and pleasure. Yet they can avoid self-denial and lean towards selfishness as they take the sensitivities and needs of others lightly. After all, a friend is a gift you give yourself. "I'd like to be their friend because they seem to really enjoy me," said no one ever.

But if you've heard the phrase *"it's not what you know, but who you know,"* you can see why a highly ranking extrovert would correlate well with success, high managerial positions, and increased income.

Questions for your potential mate:

- *What do you enjoy doing in your spare time?*

There's not much more to ask in this field. However, you want to pose the question; their answer will relate to external or internal activities. It's common for anyone to give you more than one thing they like to do, hence why it only requires one question. In the case that your date gave you only one activity, well, that might be an answer within itself (introvert), or they're just not great at conversation.

Agreeableness

As I mentioned before, it's common for someone to score high on conscientiousness and also in agreeableness. People who value religion tend to rank high on agreeableness, and most likely also value family, tradition, and established norms. While agreeable people stay away from drama, they commonly have healthy friendships since they tend to put more emphasis on others than themselves. This could be due to two reasons—they are either genuinely sympathetic towards the needs of others, or they're avoiding confrontation. Excessive agreeableness in the long term can lead to passive-aggression. So be mindful of improvement.

Unfortunately, highly agreeable people rarely have professional leadership positions, and because they care about the well-being of others, they can become victims of opportunists who take advantage of them. As mentioned earlier, Agreeableness is one of those fields that has a 60% gender variance. As a quick refresher, this means that, if you get a random male-female pair, 60% of the time, women will rank higher than men. Psychologists suspect this is due to women, on average, being more nurturing than men. You will learn more about a woman's nurturing nature ahead in our "What Men Want" chapter.

I don't have a set of questions for you in this field because agreeableness is nearly impossible to inquire about in the initial stages of love. Instead, agreeableness more closely relates to the actions the person takes. What I *can* do is give you key pointers to keep an eye out. As your conversation progresses, agreeable people usually have a nice word to say about everyone they know, as they are very empathetic and trusting, and they rarely talk about their virtues.

Neuroticism

Neuroticism is another dimension of the Big Five, where women rank higher than men with a 60% variance. There are two possible explanations for this: biological and societal. The biological reason, as Jordan Peterson explains, could be due to something entirely majestic, that a woman's biology is designed for the mother-infant dyad relation. In other words, women on average, worry more than men because their biology is intended to worry for when they have children (rationally, since an infant is exceptionally vulnerable and requires attentive care).

Another explanation for women testing higher than men in Neuroticism the majority of the time is because of the Social Role Theory in psychology, which articulates that this personality trait in women is an adaptation to imposed societal gender roles throughout history.

Neuroticism refers to a person's ability to experience negative emotions. While it's normal for a woman to test higher than a man in this field, a man who test high in neuroticism, not so much. People who score high on this trait are likely to react emotionally to stressful situations, have a more heightened sense of vulnerability, suffer from anxiety, be unstable, and aren't great with challenges. Though scoring high on Neuroticism is not to be condemned to a tragic life, engaging in introspection, adopting best practices for self-improvement, and devoting time on building their self-worth could prove highly beneficial.

Questions for your potential mate:

Though this is one of those fields that will be noticeable by the person's actions rather than chat, a great question to probe for someone's level of neuroticism is a simple variation of:

- *What do you envision yourself doing in the next five to ten years?*

Someone ranking high in Neuroticism is unlikely to be comfortable with this question and might choose to avoid it, dance around it, act defensively, or answer it vaguely. Whereas someone who ranks low in this trait would answer firmly and give you a thorough answer.

From a personal point of view, I think you may be surprised by how many people haven't put any thought into this question. A lot of people are uncomfortable thinking about the future as it is an unknown and mysterious place. Neuroticism correlates with low income and poor performance in professional fields. Thus, a person with this trait would feel discomfort thinking about the future, which in turn leads to avoidance → excuses = cognitive dissonance, a term you read about in our previous chapter and will learn in detail in the next chapter.

Is There Anything Else I Should Know?
ASSESSING THE FIVE DIMENSIONS

Yes! There is more you should know before making a decision about this lucky son of a gun. All right, so you've enjoyed a delightful evening (I hope), and from what you can tell, he seems to match up with you. Nice. But before you start picking out dresses and making a guest list, take some quick key pointers with you. *Where does intelligence fit into the picture?* And *what are the most prominent qualities for a successful relationship?*

IQ and the Five Dimensions – We established that compatibility in intelligence and personality hint towards finding an ideal mate with whom to have a loving relationship. But does intelligence and the five traits have any correlation whatsoever? To an extent, yes. Let's put it this way:

The closer the intelligence between the two →	The more compatible you are
Intelligence is primarily correlated to two traits →	Openness to Experience (Intellect) and Introversion

As previously stated, women tend to date across or up the intelligence ladder. Yet, a polarity in intelligence could prove problematic. Unfortunately, the more intelligent people are, the more cognitively and emotionally separated they are from society. If you have doubts about this, take a quick look at Einstein, Tesla, and Newton—not the best at love.

Therefore, the more equal your match is to you, the most probable you'll be happily compatible.

Secondly, while introverts don't necessarily test higher in intelligence (fluid intelligence), they do seem to process more information, which is described as crystallized intelligence. On the other hand, Openness to Experience is the one trait professionals agree on to be a good indicator of intelligence. This would translate into a personality that is average between extrovert and introvert and someone who ranks high in Openness to Experience.

Most Prominent Qualities

1. Low Neuroticism   2. High Conscientiousness

3. High Agreeableness   4. High Extroversion

Upon seeing the four qualities that studies show to be more closely affiliated with successful relationships, you shouldn't denote this as a cue to change your personality. In our next chapter, "What You Deserve," you'll learn there's always room to improve our chances at love, and while we should strive to improve some of our traits, you should not change the core of what makes you, you.

So What Now?

Should you take your results and run with the first person who best matches your five dimensions and conversely bat a home run out of anyone who seems to rank opposite of you? Well, not necessarily. As I've stressed, personality is a complex topic—to say the least. With that in mind finding the person who's right for you should be a decision that culminates all factors into one giant package—traits of a quality man (attractiveness, emotional acuity), chemistry (laughter, sex, interests), and the five dimensions of compatibility.

Simple right? I know, but I'm confident that by the end of this book, you'll have a heavy arsenal to beat this decision down with ease.

SEVEN

What You Deserve

INCREASE YOUR VALUE AND ~~SELF~~ WORTH

"
If everyone is always the problem…
Maybe the problem isn't everyone else.
– Unknown

S o, you're ready to start dating? So was I. Don't get me wrong, dating can be lots of fun…when we do it for the right reasons. But it can also be super frustrating, often leading to a lonely road when we do it for the wrong ones.

I mean, what's not to like about dating? You get to know different people, enjoy delicious food, drink some relaxing beverages, perhaps have some good laughs, good sex, and maybe a spark. See it as being open for business, taking in applications, and holding interviews. Now, if you find yourself thinking, *hmm, I don't find that at all exciting*, the issue might be that you're unsure of your business needs. Or what's worse is that instead of taking in applicants, you might be looking for a job interview yourself. It would be best if you nipped this mentality in the bud by realigning your priorities and remembering that this is all about you. Sure, in earlier chapters, we were able to define what you want. But knowing what you want for your "shop" and knowing what's essential to run it are two different things.

"Neato. Don't get me wrong, I'm all for change, but no, seriously, guys suck. There's just not much out there."

No, I hear you. Some men sure do suck—even a lot them, but there's undoubtedly someone out there for you. Overall, our success it's still on us—ownership. The problem is that we make three grave underlying mistakes:

1. **Maybe the problem isn't everyone else**—learning to look within and fix yourself first.

2. **Happiness is a state of own**—the mistake of "you make me happy."

3. **You don't get what you want, but what you deserve**—subjective perspective is not objective reality—increasing your self-worth.

It takes *vision* to know what you want, *perspective* to acquire it, *maturity* to know when you've found it, and *responsibility* to keep it. As you've become familiar with our purpose, our structure consists of recognizing the problem, redefining our perspective, and learning how to change it.

Finding the Light

As I was going through one of those rough patches of life in the world of love, I met my great friend and business partner, Sebastian, for some drinks. We hadn't met in a good while, so I kept some personal dilemmas to myself. Even so, it was nice to talk about my situation with someone I trusted, finally.

It was a rough time in my life, and most of my good friends had either moved, were now married with children, or seemed to be going in a very different direction in their lives than I was. Almost as if I was going against the current. What little advice I got when I was able to vent a little with some newer acquaintances, though well-intentioned, was full of macho "meathead" advice.

Advice such as *"You should just get laid"* or *"go live the bachelor life man, to hell with her."* Which is to say, the advice wasn't all that helpful. And I'd be lying if I said I didn't try putting some of their advice into practice. Following some of my friends' suggestions didn't fulfill me, and my experiences felt—empty. The laughs, chitchat, and drinks followed by casual sex always seemed to conclude the same way—a lack of intimacy and a desolate place by myself at the end of the night, leaving me wondering why I couldn't fill what seemed to be an unfillable void.

As the drinks rolled on that night with my friend Sebastian, and the rust disappeared from our initial small talk, we got to talking about past shenanigans, about life in general, and then relationships. Sebastian too, had started a relationship a couple of months prior (*what an excellent year for me*, I thought), and she was quite the catch. Before long, it was my call up to the stage to spill the beans about my current situation. All while I pretended to be unruffled by the subject, even though I was secretly dying to vent worse than a 1990s AC unit with no water. And so I did.

When I look back on that night, I wasn't expecting much aside from expressing my thoughts and feelings to an old friend. Much to my surprise, I received some of the most genuine advice I have ever received—with a dash of tough love, I might add. Even though it wasn't what I wanted, it was what I needed.

"You feel that way because you're doing it all wrong," he said. "You're simply following empty advice. I would have told you the opposite. First of all, entertaining the situation with your ex will only bring you more pain. Is that all you're worth? On the contrary, I would have told you to be alone. It's not what you want to hear, and it's going to suck, but you need to feel the pain until it passes." He then went on to tell me a relatable personal story of his own, after which he concluded, "You can sleep around all you want to and jump into another relationship if it pleases you. I'm not telling you how to live your life, I'm simply letting you know from personal experience that you won't ever fill that void that way, and least of all, find happiness. You have to face your fears, find them, and jump. Once you see fear directly in the eyes, run right towards it, and then, while you're in that darkness, in time, you'll soon see a light at the end of the tunnel. As you start walking toward it, you'll see the silhouette of someone you can't make out waiting at the end of it. It may take you a while to reach it. But when you do, you'll be happy to realize the person waiting all along was you. And once you meet him, you'll then know you've found yourself, and only *then* can you continue on your path."

No b.s. that is a legitimate story. I know, what a guy, right? Sorry, ladies, he's still with his pretty lady. Being that I'm terrified of heights, I soon after decided to go skydiving with a friend for the first time. So, I guess you could say I took the "find it and jump" advice quite literally. It was then that I started a journey of self-discovery and self-fulfillment, slowly learning to selfishly care more about myself by working on me, by improving me, eventually loving me, and ultimately finding happiness within me. So later, I could give the gift of all that Hugo is, to that lucky someone.

If it sounds corny, or you don't believe me, find out for yourself. Have the courage to prove yourself wrong. You'll find you have little to lose and much to gain.

It's possible you know someone single who has that mentality and fits that description. The highly independent have easily identifiable traits—

they're usually confident (not cocky), they love working on their bodies, and have been single for a minute or two. They might even be a bit selfish about sharing their time, and what's even more appealing, they might not even seem all that interested in constant casual dating. Yet they keep an open mind to finding someone special.

If you want to jump on the bandwagon, let's go, *vamonos!* You'll soon realize, as I did, the path of solitude is different than the course of loneliness. Loneliness is sad and gloomy; it's being afraid of spending time with ourselves and seeking the company of someone else to deter ourselves from the feeling.

Solitude, on the contrary, is having options and being happy with yourself until you choose to give yourself to that special someone. Loneliness is rushing, while solitude is patience. Loneliness is constant dating, while solitude is occasionally getting to know people. Loneliness is discomfort with oneself, whereas solitude is happiness, joy, and appreciation for life. If you are lucky enough to give yourself the opportunity of such a journey, nothing and no one will be able to take that away from you ever again.

Maybe the Problem Isn't Everyone Else

Upon hearing the quote, "If everyone else is always the problem, maybe the problem isn't everyone else" for the first time, someone who perfectly fits the description might immediately come to mind. What description am I referring to, you ask? You know, that person who always seems to be dating, never really changes their "single" relationship status, and when they excitedly boast about that next "great" poor someone, they're jumping into another relationship, all while condemning you to meet yet another partner, all so they can be gone before you can finish your apple martini.

On the other hand, we're often guilty of fitting such descriptions ourselves. In either case, this person is left in total shock when the briefly lived relationship comes to a sudden end, as if it was by complete surprise. Yet, they fail to see the "surprise" wasn't much of a surprise to everyone else—only themselves.

If we can extrapolate something from that, one would be safe to assume that unless this personality changes soon, they'll quickly find themselves down the same hole and repeating the same vicious cycle. It's because of that one of the key takeaways from this chapter that continues to resonate throughout the remainder of this book is the importance of avoiding behaviors and situations that lead nowhere but the self-fulfilling bottomless pit of self-loathing.

Part of the reason why you see so many failing relationships these days is that a lot of people go into relationships with a "let's see what happens" mentality. They fail to see the problem within, only to be in complete awe, wondering how it didn't work out. No need to wait—I can answer this for you right now.

When you adopt the "let's see what happens" approach, there's not much to see, so you might as well stop before you get started. Choosing this method when approaching the subject of relationships and love is most certainly the ingredient that will spoil the recipe and be met with failure every time you dive into dating with an impeccable, nearly 100% failure rate. But if no one is to blame, well, precisely, *who do we blame?* I think you're hating knowing the answer already.

A relationship can be many wonderful things. It can be fun, create joyful memories, and bring the best out in us. Nonetheless, we must understand—and what's more, *accept*—that anything worth having in life takes work. I'm not referring to the work in the relationship you might be thinking about.

Sure, a relationship requires work, among other things, but that comes later. For now, we'll focus on working on ourselves first. We far too often make the crucial mistake of becoming fixated on the "good stuff"—the laughter, the outings, the trips, the eye-rolling sex, and all the great little things that come with a relationship. Hey, I get it, it's hard to blame you. Get me some good, ol' heart-throbbing orgasms, mouthwatering desserts, and don't forget the wine while we're at it.

Perhaps somewhere through our lifespans, we heard the phrase "can't be all work and no play," and boy, did we run with it. So, I say to you with a heart full of empathy; *it can't be all play and no work*, either. Any potential healthy relationship will require hard work at times, regardless of how much we look forward to leisure.

Besides, saying a healthy relationship requires "hard work" doesn't mean that fun and work should be of equal balance or that you should hold a pessimistic view of what work actually implies in a relationship. On the contrary, work amplifies the leisure.

Have you ever gotten off from a long day's work and sat back and relaxed to watch some TV, and felt like it was the best thing in the world? Or have you ever come out of a common-cold and been comically grateful to be able to breathe normally again, and not have a headache? Whatever your experience has been, most of us can relate to one or another moment in which a temporary struggle has amplified our appreciation for the reward. It is work that magnifies the results, and in the same essence, as opposed to common belief, it is the consistency of "all play" that can often lead to complacency.

Once, while talking about this subject, someone told me, "I love being on my own. I mean, I'm independent as f***." Regretfully, I didn't have the heart to tell her at the time how incredibly wrong she was. What she was,

was lonely. She fit the description in the paragraphs above nearly to perfection. She jumped from one relationship to another, never held a meaningful, lengthy relationship, sought the constant company of friends, and dated more guys than she could count.

Many of us are guilty of lying to ourselves as well or have in the past, myself included. The important thing is recognizing it. We can't do anything about our past but focus on the present and learn from it. Stop lying to your reflection and self-reflect. To be able to recognize the issue within, work, improve, and move forward, you must take ownership of the pen that writes your story.

What could be cynical about that? And if it's that simple, why haven't we done it already? I hate to be the bearer of bad news, but the bitter pill to swallow is, realizing we have accepted our failure, and rather than seeking rectification, we conform with justification. That's right, as long as someone *sees what happens,* they can never be held accountable for their failures. Almost everyone has a friend who, *God knows*, nothing is ever their fault, and life is always happening to *them*, instead of the obvious—life is only *happening*. Besides the fact that it's utterly irresponsible. People's feelings are not to be played with.

If we applied that roll-of-the-dice, cavalier, nonchalant mentality to anything else, people would immediately dismiss it. For the sake of example, if I said I'm buying a pup, but I'm not sure if I want to keep it, I'm "just seeing what happens," or upon opening a new business, I didn't have a goal and just wanted to see where it goes, it would sound ludicrous. You'd want to slap me across the face to knock some sense to me, and rightfully so. You'd consider me highly irresponsible. Nevertheless, people do this when entering relationships *all the time!* So, if you don't want to work on yourself, learn about dating, for that possible exciting relationship, then you shouldn't be in a relationship, period.

If you think humans are different because a pup's love is unconditional, whereas a human's love is not, you're right. The purpose of the example was to provide perspective into the irresponsibility of our vision and thought process. Still, in any given case, no one's dismissing the possibility of failure. But failing because of the *fear* of failing is as ludicrous as not taking a step forward out of fear of tripping. Failure is always an option in any aspect of our lives, but not risking failure is not risking success either.

"The road to success is paved with failures."
Rashida Rowe

Cognitive Dissonance

The truth of the matter is that entering into a relationship is as simple as making a decision. Yet, the thing about free will is that despite being privileged with the power to choose, making a choice can be, well, freaking hard. Hence, why it's simple but not necessarily easy.

When I say that something is simple but not easy, what do you think that means? It means that while the idea or understanding of a concept can be simple, putting it into practice is a different thing. It's an interesting philosophy that applies to an array of aspects of life, including— relationships. While this philosophy doesn't tell us how to solve a given problem, it does lend us perspective, which allows us to view the topic from another angle to understand it better. Take the following image into perspective:

A   --- >   B

100 Miles

If I told you that to reach your goal, you have to travel a hundred miles, but the only way to achieve it is by foot, how hard would it be to visualize the illustration above? Well, I hope it wouldn't be *too* hard since you just did. It doesn't take more than looking at it for your brain to upload the image into your mind. But it's here where things go awry for most people. Why is that, do you think? That's because even though visualizing an action is part of the process of progress, experiencing the discomfort associated with literally walking those hundred miles is a physically taxing task since it requires actual energy.

Enter cognitive dissonance. *Oh, boy, here we go*—time to turn on the lightbulb. Cognitive dissonance is when new information is different from your beliefs, intellect, or personality, ultimately creating inner conflict. This conflict presents itself as discomfort, and heaven knows, nobody likes that. *Take my health, let me break some hearts, but for goodness sake, anything but discomfort!*

It's in this moment where the ugly duckling, antagonist, the lazy cousin to reason, joins the show and wrecks—sorry, saves the day, by reducing the dissonance. Hence, *excuses.* The whole thing looks something like this:

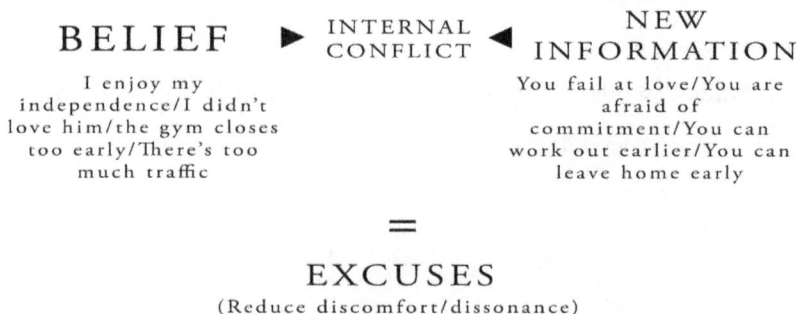

BELIEF	▶ INTERNAL CONFLICT ◀	NEW INFORMATION
I enjoy my independence/I didn't love him/the gym closes too early/There's too much traffic		You fail at love/You are afraid of commitment/You can work out earlier/You can leave home early

=

EXCUSES
(Reduce discomfort/dissonance)

Timothy A. Pychyl, Ph.D., a faculty member in the Department of Psychology at Carleton and writer for Psychologytoday.com, tells us how one of his student's dissertation research shows that people usually fall into one of four preference structures. She lists them from the most chosen to the least:

Rationalize away the behavior (ex. "It doesn't matter what I do, it won't change the outcome.")

Deny responsibility for our behavior (ex. "It's not my role to do this, they failed me, I lived with him.")

Distract ourselves from the dissonance itself (ex. "I've got other things to think about right now, my job is demanding, I have kids, I just don't lose weight.") Procrastination.

Change our behavior (ex. "I will take the time right now to address this issue and improve."). ← Said no one ever. Okay, maybe a small percentage, but they are a small minority.

We overcomplicate the world of love because we live life according to a formula we didn't even know existed, let alone that we were using. Well, you are! It is the cognitive dissonance formula. The problem is not that you are using a formula, but that you're using the wrong one. It's much easier to say you don't know what you want, and once it fails, not feel as bad about the outcome than fail after expressing how important something is to you and giving full effort. But is it? I happen to disagree, and many experts would side with my disagreement. The truth is, there isn't substantial data to even prove our excuses reduce our dissonance level. Be that as it may, I believe the problem lies in our perception of *failure* and our *understanding* of reducing conflict and stress.

In the utmost simplicity, we tend to think of failure as a permanent state of affairs when it isn't. The only certainty and permanent state in life is death. What a cliché, right? I'm not joking—it's a simple reality. Failure is not even close to being permanent. Instead, we should see failure as a learning opportunity because it is. It's imperative to understand and accept that failure is a next-door neighbor to success, and they say hi to each other in the mornings. Therefore, if we can't change the processes of life, we may as well change our perspective and hence, ourselves.

When we succumb to the more comfortable and often traveled path of excuses (which we call 'reasons'), we do it out of the assumption it will alleviate our discomfort (dissonance). Yet, it will only reduce it, yet never completely get rid of the discomfort. That's because once you have received new information, the inception has been made, and the mind is really good at reminding you of pain. If you're willing to accept this as truth, the only way is forward. The only way is to confront yourself because, in the long run, you'll cause more harm than good. But I know you're on the right track, or otherwise, why would you be reading this book?

After holding yourself accountable, we can move forward and answer the following questions: *What am I doing wrong? How do I change my behaviors to avoid the path of loneliness?* In short…we find *happiness* for the beautiful lady in the mirror first. How? We help her remember how to value herself and how to increase her worth, so the quality man she wants in her life perceives it.

Happiness Is A State of Own

In the Declaration of Independence, "the pursuit of happiness" is defined as a fundamental right to freely pursue joy and live life in a way that makes you happy, without violating the rights of others. To better understand its meaning, let's break the context down a bit.

One of the first takeaways that immediately pops into mind upon hearing such a definition is that happiness is a pursuit. This idea implies that, if we're pursuing happiness, then it must mean we don't have it to begin with.

While most of us think of happiness as something to be acquired by chasing or searching externally, instead, the pursuit of happiness refers to the *experience* of happiness, the *practice* of happiness, and *finding* it in our daily lives. In a 1964 essay titled "The Lost Meaning of 'The Pursuit of Happiness'" by Arthur M. Schlesinger, the author does a great job of shedding light on this point and redefining the social perception of the term.

Thus, happiness is a state of *own*. We must find happiness within ourselves first, and then the people we choose to allow into our lives will emphasize it but will never be able to provide us happiness. If we don't address the issue with the courage to look inward, the tape on the leak will need to be continuously replaced.

If you think you're not courageous because you're afraid of change and looking inward, you couldn't be more wrong. We're all afraid! There's no such thing as fearless individuals, only courageous ones appearing fearless. Courage is not being unafraid but *being afraid* and finding the courage to face it. It takes courage to look inward and be brutally honest with oneself since a lie to the mirror will always be more comfortable and hide as a subtle "reason" pointing elsewhere. Conversely, if you are honest, allow yourself to heal, and learn, you will inevitably fall in love with your reflection, and when that moment comes, then you can give the gift of you to someone who's done the same.

Finding Balance

Everyone can benefit from spending a bit more time with themselves. Ideally, spending enough time with yourself where you reach a healthy level of comfort. Make no mistake about it, we are social beings by nature. Human beings cannot live sanely or healthily without social interaction. So I'm not talking about secluding yourself in a dark, quiet room indefinitely by any means.

There's no exception to the rule. Our minds simply cannot live without social stimuli. It doesn't take more than looking at some of history's greatest minds, Nikola Tesla and Isaac Newton, for example. Both were geniuses among geniuses who preached and practiced complete isolation yet suffered from depression and almost went mad. I'm sure we can be thankful for their sacrifice and contribution to our lives. But in your case and mine, unless we're in the process of redefining mathematics or inventing tomorrow's energy source, I'm confident we can take a raincheck from isolation.

Teaching and training our brains to self-soothe is very much an art. To an extent, stimulus gained from socializing can be obtained from something as simple as watching your favorite show. Your brain is not all that great at telling the difference, just like it's not great at differentiating porn from reality, in turn affecting both a woman's and a man's performance. Still, chapter fourteen is waiting ahead with a lot more detail.

Remember the scenario with my "I'm independent as f***" friend, yet her actions highlighted with orange marker how dependent she was? Well, dependency is what we're trying to alleviate. No one's telling you to become a monk. All we're trying to accomplish is reaching a real state of independence where you're happy and comfortable when you're by yourself, where your home is a place of Zen, where you make yourself smile, often not even understanding why.

Our goal is to find a balance between the polarity of emotions, thoughts, and in this case, comfort to reduce our dependence levels, and that couldn't be healthier. Our goal is to arrive at a place where you understand who you truly are, ideally reaching the pinnacle of independence where bringing someone into your life is not out of necessity but a conscious and amiable choice.

You might remember the term Cognitive Behavioral Therapy (CBT) from our breakup chapters. As a slight refresher, CBT is an evidence-based psychological treatment mainly used for anxiety disorders, but it's also proven to be highly efficient in different situations of psychological distress.

One of the great proven practices of CBT is known as Exposure Therapy, which involves slowly acquainting (exposing) the patient to those objects or situations that provoke fear to confront that fear or anxiety slowly and efficiently. Now, it is vital to make it clear that you should never self-diagnose if you think you may suffer from a medical condition. If you find yourself facing severe anxiety issues or symptoms, you should consult with a licensed professional. This lecture intends to help people in their love life and everyday life, but it is *not* professional advice for anyone suffering from a medical condition.

Overall, we should aim to alleviate three main things: *dishonesty, fear,* and *dependency.* Dishonesty with ourselves, stopping the lies we tell ourselves out of fear. However, something tells me you're doing pretty well with putting that into practice since you've continued reading this far.

Secondly, let's work on ridding ourselves from the fear of being alone because there's absolutely nothing to fear as long we acquaint ourselves with who we are. Otherwise, it's that fear that leads us to dependency, the third anchor of which we're trying to rid ourselves.

How? is always a great question. The fascinating news is that with a couple of tweaks here and a few adjustments there, we can tackle all three at once. Hooray! Drinks on me (not really).

Buying a daily planner will facilitate your whole routine and make the process run swiftly. I couldn't stress enough what a beneficial addition it will prove to be in your life. Aside from that, a high starting point would be to continue the routines from our "coping" breakup chapter. But not to worry, I've brought over the ones that will best benefit you on this point onward in your journey with some slight modifications—*Exercise, express, eating healthy, meditating, activities and hobbies, goal setting, and traveling*. You can take it to the next level now.

If you've passed the grieving stage of your breakup, set a body goal, continued with your method of expression, set a new health standard— lower blood pressure, try out the vegan life, or perhaps consume more veggies. If you dig mindfulness meditation, by all means, stick with it, but know you can become one with yourself with several different styles of meditation. Take yourself to a nice bar or restaurant or both. Take yourself to a movie you've wanted to see; Sunday matinees were my favorite self-treats for a long time. Continue working on your set goal, or set a new challenge for yourself if you've accomplished that one.

As you may have noticed, I added traveling. We tend to think of traveling as a social affair. I have a good friend who loves to travel, and she won't let anything stop her. She takes herself all around the world, as she's already in the best company. Traveling can change your perspective in life. Paris may awaken the artist in you, while Dubai may make you aspire for a more lavish life. But traveling can also make us appreciate the small luxuries we have that go unnoticed, as seeing other lives, cultures, and ways of thinking opens your perspective on the world.

When you are in the comfort of your home, ask yourself what entertains you when at you're home. Puzzles, watching TV, listening to music, reading, working, games? Figure it out. The first step is asking the question and engaging your thinking power to find an answer. Once you've planted the seed, the solution will ensue, so don't sweat it too much.

Remember that complete isolation is not at all healthy. Take your rear outside and now ask yourself the same question about the outdoors. Is it walking, biking, jogging, a drink, a meal, a patio atmosphere while reading a book, some tasty coffee, a movie? The truth is we have infinite options. So many that if we wanted, we could write a whole other book on the list of options. So, you have *zero* excuses. Life is plentiful. Take yourself out to greet it all while reminding yourself of the truth: you are not alone; you are choosing to spend time with yourself—because you want to make better decisions upon those options.

The same applies to pets. If you already have a furry companion, great! If you don't, and it was recommended by a medical professional to deal with any of many diagnoses, whether it be anxiety or depression, fantastic. Follow through with such advice. If you have reached your balance, independence and have found happiness, incredible! Get that cute, little, love furball.

Otherwise, though a pet can be a beautiful addition of unconditional love, warmth, and energy to your home, I suggest waiting until you reach or encounter any of the scenarios mentioned earlier. Please don't misinterpret my point; pets are unbelievable, nearly too good for us to deserve. Not only that, but even science cheers them on and supports the incredible benefits of their company.

Nonetheless, for one, a pet is a live being of Mother Earth that requires serious responsibility. Like any other significant responsibility, it should be taken with thoughtful consideration and logical reasoning. Getting a furry friend to compensate or soothe our emotional cravings is hardly a justifiable reason. So, meanwhile, let's stick to the plan. You're strong. You got this.

We Get What We Deserve

One of the key takeaways from the previous sections in this chapter is the "total shock" some people experience when they're suffering from their dating life or other times when the relationship didn't take off. You might be well-acquainted personally with common grievances about the opposite sex, whether it be either through conversations, friends' comments, or posts from social media. For the sake of the point, let's imagine "Linda" for a moment. Linda rants about how in one way or another, *all men* suck.

If anything should be shocking at all, it's that "Linda" is shocked in the first place. It's like showing up to the dealership with thirty grand and wanting to buy a Ferrari, and then, what's worse, being shocked that you didn't walk out with your bright red sports car. What? No, Linda, no! What our fictional Linda here fails to understand is that to get that ultimate sports car, she has to *earn it*. In the fun and confusing world of dating, she has to elevate her worth. Otherwise, Linda better get used to her brand-new Maxima real soon, if that. So what's wrong with a Maxima? Absolutely nothing. My editor reminded me that "Maximas are the s****!" What's wrong is Linda's set expectations.

Don't get me wrong. Linda might also be right about all men sucking— in *her* reach circle, not all men in actual existence. Let me explain. This is called the ripple effect:

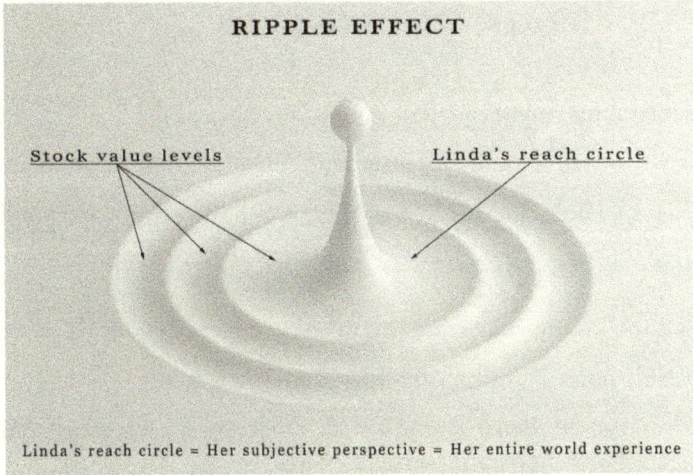

RIPPLE EFFECT

Stock value levels

Linda's reach circle

Linda's reach circle = Her subjective perspective = Her entire world experience

There's a couple of things to understand here. Now, if you dropped a rock into the water, you would get a ripple effect; the heavier the rock, the farther the ripple will reach, simple. Likewise, the heavier your worth, the more you amplify your reach circle. An essential fact to understand is that subjective perspective is not objective reality, which helps us explain the fascinating spell from which Linda suffers. In other words, our perspective formed by our personal experiences is *not* the reality of the entire world. Yet, it is *our* whole world, hence becoming *our* reality.

But what does this mean exactly? It means that Linda's frustration towards men is probably a frustration towards herself. Now, what Linda must first do if she's genuinely interested in better candidates is to realize this analogy. She must amplify her reach circle by increasing her worth. Our options are a complete consequence of our value. It's also important to note that, of course, this doesn't mean that by amplifying your reach, everyone you meet will be Prince Charming. It means you can now enter the jewelry store's private client foyer to view the better options you've wanted all along.

Increase Your ~~Self~~ Worth

Once you come to this realization, you've officially opened the doors to new potential. As we move forward, we must continue with an open mind and remember this is a book, and because of that, it is limited to broad points and generalizations. Besides, honesty is a meal better served cold.

So, what does "raising your value" mean? As you know, one of the major highlights of this book is realization, which is only a result of practicing sincerity with ourselves. Being truthful with ourselves is a necessity. The simple truth is if we're not attracting the type of person we envision in our lives, we're probably not at the level.

But into what does this translate? It means that if women look for physical attraction, financial status, and masculinity, among other qualities, a man should improve himself in those areas to increase his probability of success in finding the partner he envisions.

From the time I was twenty-four, I've made decent money, especially being single with no kids. Even though I was dating, there were still women I was attracted to who weren't necessarily attracted to me, and I simply couldn't understand why. After all, some of those women dated guys I didn't consider impressive in any sense of the word.

They weren't particularly good-looking, and they weren't the sharpest tools in the shed. I noticed how these women would always complain about how these guys treated them. So for the life of me, I couldn't figure out why I wasn't the one dating these ladies. I consider myself to be a good-looking guy, I'm financially stable, I'm no Albert Einstein, but I'm pretty intelligent, and yet all those guys had something over me—money. So I guess there's nothing I can do about that and should resort to complaining, right?

Of course not, I simply accepted the reality. I either *conformed* to my current standards, or I *changed* my circumstances. I'm naturally very competitive, so I went with the latter. Today, I own two businesses, work from home, and am now an author. Sure enough, my dating life significantly improved.

As I've mentioned before, I went through one of those rough breakups when I was twenty-eight. Once I felt ready to date again, I was surprised and humbled to see so many nice and beautiful women were attracted to me. I'm no Casanova, and though I'm no angel either, my dating life had never met that level of success. That's why I'm confident there are plenty of good men waiting for you, whenever *you're* ready to let them see your shine. But there's no fire—they can wait. For now, let's take it a step at a time and work on shaping that diamond in the mirror the look *you* want. You set the pace.

As we reach the end of this chapter, we have learned that worth and happiness belong to us. Furthermore, when we empower ourselves by taking ownership of our lives, we can directly influence our outcomes in this complicated world of love. I'm excited for you, what's to come, and what's to become of your future relationship. If you're ready to embark on that journey of relationships, let's continue our trip to the chapter ahead and enter the men's chamber of secrets if you're curious to know what men think.

EIGHT

What Men Think

MEN'S CHAMBER OF SECRETS

"

*Do you solemnly swear to tell the truth, the whole truth, and
nothing but the truth, so help you God?*

— Unknown

Ladies, if you only knew what you have in your hands right now,
you'd be setting this book on top of your bookshelf. Men may
hate it, you may find it hard to believe, and I may just have to
move to the south of France and change my name to Vincent. This chapter
will break so many rules of the infamous "guy code" it's not even funny; it
might also be illegal, for all I know. Some things you'll love to hear, others
you'll hate and may strike a nerve; some will chop at your pride, and others
you will celebrate. If you're a man, the same is true, and if you happen to be
reading this next to your lady friend—*I suggest you move.*

If the truth sets you free, then let's tell the truth indeed. *Yes*—men pee
in the shower, but so do you, and word has it, some of you don't wash your
legs, but your secret's safe with me. Men also pee while sitting down when
they take a deuce. We've all jammed to "Killing Me Softly" by Roberta Flack
or sang some good ol' Backstreet Boys at the top of our lungs because no one

said good music was only for women. We, too, cry at movies when alone, because we also—feel. We have plenty of insecurities too, and even though tall women can be fatally attractive, they intimidate a lot of men (I plead the fifth). And don't let men tell you they can't recognize another man as attractive.

On the other hand, we're in complete agreement that Spandex is the *greatest* invention known to man, so please keep wearing it, and yes, we'll surely be "badgering the witness" five minutes after you leave home. Some of us—I mean, them—still watch anime, and at one point or another, we have shaped ourselves a funky shampoo hairdo and gone over to the mirror to check it out, and if you find yourself asking why, the answer is: we're still kids inside, let us be. But perhaps one of the biggest secrets of all time is one you'd prefer not to hear (but you can't help but be curious). Men, too, fake it in bed. *What-what?*

Why did you need to know that? Because you must understand the world of your testosterone-driven male counterparts since that's who you'll be dating.

So men want sex. Not much of a secret, huh? But the real question is *how much?* If you know, you might not be so quick to call every friendly chap you've encountered a "friend"; you'd think twice about sleeping with us on the first night, and you might not want your boyfriend to be so close with your female besties. If you knew, it'd be more likely for you to be staring blankly into space while sitting on a swing.

But that can't possibly be it. So aside from all the female body parts lying around a man's brain, what do men really want? What do they like? And what do they hate? If you can't wait to know, here's your one-time backstage pass—proceed with caution (really).

Men & Sex

When navigating the world of online dating, it doesn't take more than two or three swipes before you bump into the standard barrage of comments, such as *if you're not going to respond, don't waste my time* or *if you're looking for sex please swipe left.* I hate to break it to you, but if you thought you were going to meet a man who is not looking for sex, respectfully, kindly, I'm afraid you're idealistic—at best. Also, a man who tells you that he is, is—pardon my French—full of shit.

Now your immediate reaction might be, *Well, you're just saying that from your personal experience. Not all men are like you.* Okay, Karen, duly noted. The truth is, sex is in a man's biological nature; therefore, sex is **always** a man's intention, without exceptions, and the faster you accept this, the better equipped you'll be for the men out in the real world. I hear your protest though; it's not what you want to hear. But what you might really mean is, it shouldn't be a man's *sole* intention as there is more to you than your body and sexuality, and perhaps you also expect that even when it is our sole intention, we shouldn't be so darn corny about our approach—*Hey, you come here often?* By all means, you are correct in not wanting to be seen purely as someone's favorite meal. Therefore, how do we solve this conundrum for the sake of better future relationships?

Well, here are a few useful pointers. If men happen not to respond to you—ghost you—there's no need to overthink it, since it has nothing to do with you. It's as simple as maybe he was busy. Perhaps he met the love of his life and closed the apps, or unfortunately, he had a different personal preference. Whatever the case may be, the truth is that whoever is interested in us will take the time to respond to us. There's no need to invest negative energy in a situation that deserves no more of your precious time than what you've already spent. It's probably not what you want to hear, but you shouldn't let it even remotely affect you. Expecting everyone to whom we're attracted to like us back is subconsciously narcissistic. It is our brain saying, *how dare they not like me? Don't they see what they're missing out on?* You might be a hundred percent correct, but the point remains, not everyone has to like us, nor will they.

Besides, try to see the lighter side of things with a bit of humor. If you think a couple of guys here and there not responding is bad, I recommend you take a look over at the male side of dating. Just ask one of your male friends if you can peek at his online dating apps and then count how many times women don't respond to him. I'm sure not only will you view your situation differently, but you may also get a kick out of it.

As we move forward in our text, it's natural for certain things you read to strike a nerve, but please understand that's not the intention. So sure, men primarily want sex, and we want it a *lot*. Yet that doesn't necessarily mean all sex men have is good either, as I'm sure you'd agree. It only means that in general, we think of it and want it more than you, that's it. Therefore, try to keep an open mind as you read and not be so quick to adverse criticism. Ultimately, women and men are two sides of the same coin, and I'm merely showing you our side, while both of us try to make things work out.

Back to sex—I, for example, have been at both ends of the spectrum. I've had sex that was so atrocious, I found myself like Ralph from *The Simpsons* thinking, "I'm in danger!" Other times, for some reason I'm not able to dissect, it was me who wasn't able to perform. While all men seek sex, like women, men also have moments where we don't feel like having it. Whether we're tired or simply don't feel turned on at the moment, it's normal. Yet, one of the frustrating aspects of that fact from a man's perspective is that when a man's libido is absent for whatever reason, it's more easily chastised. Ultimately, men inherit the responsibility in both scenarios.

Still, the question remains, how much do men think about sex? The truth of the matter is that nearly everything surrounding a man's life is about sex. It may sound farfetched or even make you cringe, but it's the truth. Hell, even when we're not trying to think of sex, we think of sex. If you ask us why we likely aren't able to give you an answer. The truth is, we don't know. We just know that we like uploading female body parts into our imagination whenever we see them. Otherwise, restaurants with pretty ladies and revealing attire wouldn't exist, yet men keep them nice and profitable. Maybe it's as simple an answer to the riddle as we are programmed that way.

Some of you reading may think, "Hey, I love sex too," and while that is true, we still have to dive in a little bit deeper for you to grasp better the depth of a man's mind. Sure, we both masturbate and get turned on thinking about sex. You read erotica and occasionally watch some ethical porn, and we watch porn, period. But there's much more to tell.

In 1615 Galileo Galilei wrote a letter to the Grand Duchess Christina Lorraine of Tuscany with the phrase, "The Bible shows the way to go to heaven, not the way the heavens go." What he was trying to explain by quoting Cardinal Caesar Baronius was simply that the Bible lectures on the acceptable practices to get to heaven; it does not explain what the heavens are. Galileo was scientific in his approach, of course, and meant that the teachings of the Bible and the studies about heaven should remain separate affairs.

Using Galileo's letter as an example, this text has given you advice on how to get to "the relationship" and get a quality man thus far, but not what the man is or wants; two ideas that are also separate affairs. If you're still curious, strap in, as it'll be a bumpy ride.

How Much Do Men Think About Sex?

Well, let's just say it's surprising that an fMRI scan of a man's brain doesn't spell out the word S-E-X. But a man's intention is not only sex but *primarily* sex. Now, while women may have a different definition of what an attractive woman is than men, men have only two.

Men tend to think in terms of *I-would-date-her-purely-from-her-looks* attractive and *I-would-sleep-with-her attractive*. I'm sure plenty of women have classified men of dating vs. casual sex as well, but the perspective is hardly the same. Men would be completely okay with having *only* casual sex with someone they find attractive. Women, on the other hand, might say the same, but in reality, they run a higher risk of building emotions towards their sexual partner in most cases. Despite the aggravation (I hope not too much), whichever way you spin it, sex and attraction between women and men are just **not** the same.

As I said, both you and I might have different opinions of what an attractive lady looks like. But try to put your dude hats on for a couple of moments and imagine you're a guy hanging out with some good friends at a typical sports bar or brewery. A woman passes by that a guy finds reasonably attractive. He may not have a desire to get to know her personally—he may want to sleep with her because she checks his physical preference boxes—well, that's what I-would-sleep-with-her-attractive looks like to man.

Men have an internal pair of X-ray-vision goggles. I hate to say it, but you may have even been part of their use at one point. But just because sex habituates in our minds all the time doesn't mean that's the only thing we want; only the primary thing we want.

But why should any of this hold value to you? Was it hard to read, or did it offend you or make you uncomfortable? If you answered yes to any of those questions, perhaps it should lend you perspective to how foreign a man's mind still is for some. Additionally, how can you expect to deal with these sexual mammals roaming out there? Besides, if you think that was harsh, be thankful for my editor's voice of reason, who burned half of it.

In the world of dating, it's because of these harsh realities that men hate your "guy friends." Because they think like those male friends of yours do and are relatively telepathic when it comes to reading their thoughts or intentions.

Female and Male Friendships

Can a non-platonic male and female friendship exist? Sure, but not in the context commonly expressed. Hence why male partners are not the biggest fans of your "guy friends."

Put it this way, as if it wasn't already engraved in your mind, men primarily want—? Sex. After reading the previous chapters, we also know that attraction is a complex subject. So, why would any heterosexual male want a non-platonic friendship with a woman that fits into the categories of attraction? They most likely wouldn't.

Now, no one is saying that men will proactively seek sleeping with their female friends, nor that a man would take advantage of the opportunity if it presented itself, and least of all, that you can't have male friends. I'm merely highlighting how a man's brain works. Of course, women and men having platonic and even non-platonic friendships.

The primary reasons for a non-platonic friendship to exist are—a *disparity in the spectrum of attraction* or *meeting possible consequences.* Meaning they either have very different personalities for a relationship to work, or perhaps there's a gap between the two in their level of intelligence, or maybe a noticeable difference in their physical attractiveness.

On the other hand, perhaps those gaps don't exist, but there's a conflict of interest such as a friendship that spurred from that person dating a good friend. Or the other way around, like a partner's close friend. But even in those cases, it takes a significant level of *maturity*, a genuine sense of *realism,* and a set of *boundaries* that must be set. Sadly, in a lot of cases, people are not sufficiently mature, realistic, or capable of setting boundaries. Without those elements in place, a male and female relationship is no more than—an illusion.

Respectfully, this illusion is mostly perpetrated by women, who usually rebut with clichés such as "we've gone drinking plenty of times, and he's never attempted anything" or "he's my friend's boyfriend, of course, nothing would ever happen." Yeah, best pals ever, right?—What? Whatever the case, kindly, please refer to the examples in the paragraph above. The possibility of attraction does not necessarily equate the probability of something happening between you, though it does influence it, given an array of different factors.

But, in any given situation for a man, if circumstances aligned, and a sexual opportunity was present, it would take a lot more than trust and the illusion of friendship to hold himself back from not having sex. The reason why this possibility is not to be overthought is a logical one—men sexualize all physically attractive women in their mind, and women lie to themselves or are simply naïve about a man's intentions. As said above, it would take more like maturity, realizations, boundaries, and a whole lot to lose, combined.

In a lot of cases though, that level of conscientiousness is thrown out the window and substituted with what we mentioned earlier, the illusion of friendship with a long tail of excuses. It's precisely this mentality that drives men bananas, yet one they conveniently use to their advantage when their cute female friend is in town to visit. Hey, they're childhood friends after all, right? This illusion is as dangerous as driving blindfolded on the highway during rush hour.

If this notion is false, why do coworkers often end up sleeping together? That's how chatting, and innocent joking turns into laughter, which creates more in-depth conversations, out of which they notice common interests. Does this ring a bell? Add constant exposure, shared similar traits, and you have none other than—chemistry. That's a potent mix that can very well turn into an actual *mix* of drinks at night, music, and then a "Oh, what fun— Gosh, he looked at my lips, what's that about?—And I don't know what happened, and I can't remember a thing" *moment.*

A point of no return that, in most situations, can be avoided by merely practicing some honesty with ourselves, our partner, and friends. Because without that, being your best friend's boyfriend or girlfriend alone will not suffice. *Gosh, what kind of a friend are you?* Yes, I know, but please don't be so quick to judge. After all, it was Carl Jung who presented us with the idea of shadow psychology, which says that a sensed *personal* inferiority is seen as an apparent moral deficiency in others.

Do you remember Gina's boyfriend, Tommy? Well, he remembers you and certainly reflects how you looked in that swimsuit the day you all hung out with some friends at the beach. Which is the same reason Tommy apologetically opened the door without a shirt on that day you came over to visit Gina.

It doesn't even have to be visual. Men were blissfully imagining snippets of your narrations in that "adult talk" you were all having among friends and couples. If you mentioned undergarments—they visualized it. When you comment about a private part of your body, such as saying, *"Yeah, I fell off my bike the other day and scraped my inner thigh, it really hurts"*—they visualized it. You mentioned any word with a potentially double entendre, such as swallow, wet, slimy, tight, big, or any phrase you can add "that's what he/she said" to it? While everyone laughed at it, and you thought it was funny—you got it, they visualized it and uploaded it to his imagination. I know it's harsh, it's ugly, it's nasty but ultimately *real*.

Men and women are not all that different in our competitive nature. Make no mistake about it; we ultimately want what's best for us and are not as goody-two-shoes as you'd like to think. Even a friend of the same sex is a gift we give to ourselves. Consciousness without proactive, conscious thinking and practiced awareness by inward questioning is no more useful than your programmed intelligent speaker giving the appearance of real intelligence.

Prosocial Behavior

Engaging intellect and proactive conscious thinking to be able to participate in daily acts without expecting compensation are known as prosocial behavior (PSB). Practicing prosocial behavior is not an easy task. It requires high social awareness and high cognitive power, but learning its meaning is a useful starting point, at least. In other words, prosocial behavior tells us that most of the things we do, we do out of the expectation of external or internal stimuli or to be compensated. This compensation can be so inconspicuous that we don't even know it exists.

For example, in a mice experiment conducted to gauge empathy, two mice are placed in a cage. One is constrained to a smaller compartment or tube from which the free mouse can open to liberate its partner. At first, the free mouse is not given food until he manages to free his companion.

After the mouse has been conditioned with the expectation of food upon opening the lever for his partner, the experiment now shifts to giving the free mouse food while the other mouse remained imprisoned. During the first attempts, the mouse ate his food without ever minding the entrapped mouse. Upon smelling the food and seeing the free mouse eat, the trapped mouse's stress level rose, and it became more agitated.

Surprisingly, after a while, the free mouse still approached the lever to free his captive friend. Neat, right? So was the free mouse showing empathy after all? Well, I hate to break it to you, but not necessarily. The researchers propose that the agitated state of the trapped mouse interrupted the free mouse's peaceful eating, causing him to stress out as well and finally free his friend. This tells us that the mouse was never empathetic towards the trapped mouse; instead, he was simply alleviating his own stress, which further emphasizes our previous point of self-interest and our misplaced notion of friendships.

It also bears the question, do we give money to the homeless man at the corner to genuinely help them, or do we do it for the sake of subconsciously alleviating our guilt or discomfort? Given the examples, I challenge you to ask yourself, "How then does a male benefit from seeking your friendship?"

The prosocial behavior formula would look something like this:

Engaged Conscious Thinking (Reflection) → **Awareness + Maturity** (a sense of self) → **Realization** (I am a man, and he/she is attractive) + **Consequences** (Breaking those boundaries could *lose* me my fiancé/marriage, future, kids, home) → **Decision** (Boundaries must exist).

=

Successful male/female friendship

Now, as mentioned before, life operates in constant opposites: life and death, success and failure, ignorance and knowledge, good and evil. You name it, the list goes on and on. Overall, it's the essence of shadow psychology and the philosophy of yin and yang.

So rather than take the above scenarios literally, try to see them as possibilities that become ever more probable with the confluence of unrealistic friendship expectations and the innate instinct of men's thinking.

Since we all have different lives with an array of different situations, it's all about finding balance or finding the middle ground between the polarity of life's many facets. Because just like a man's instinctive sexual thinking doesn't necessarily mean he'll act upon it, your friend enjoying your boyfriend's attention to her comments doesn't mean those are her intentions, either. Therefore, perhaps boundaries of respect should be set and met so that those thresholds of freedom between our partners and friends exist within those limits and not as often. And while you shouldn't form delusions of persecution, perhaps you shouldn't be as blindly trusting either.

This chapter should merely serve as a reminder that we are living organisms and just another animal among thousands of others in the animal kingdom. Because of that, we have programmed natural instincts as well, despite our intellectual power and being the apex predator. We have social instincts that make us want to be liked by flirting, crying to elicit empathy, by our laughter, and perhaps, our instinct to love and want to be loved and to nurture and be nurtured.

If we are to take something meaningful from this lecture, it is that in so many ways, we are indeed similar to the rest of all animal life, in that we also follow biological patterns. Still, we are also different in our privilege of consciousness, giving us the gift of choice and reason. We are the only living being equipped with the tool to gauge those instincts by engaging awareness, proactive thinking, and the power to choose.

My Zira

As you know, after being in a relationship for four years, I remained single for about two years, mainly practicing the majority of the things you see in this book. I wanted to grow, define what I really wanted, learn more about women, and fix myself to become a better version of the man I was before.

Because, if I may be frank with you, I wasn't the best boyfriend in the world, and though to this day I think we weren't all that compatible, I still believe I could've been more appreciative of the genuine love that person gave me. I grew accustomed to compliments from people about my intelligence and her regular flattery, I grew delusions of grandeur and became complacent in my condescendence. Nonetheless, I always thought I saw so much. Therefore, when we broke up, I wanted to be strong enough to hold that high-powered perception upon myself.

The months passed by, and the darkness faded and eventually dissipated. Slowly time moved forward, and the warm breeze turned cold. I began dating, and with every face, I learned a different story, learning a bit more about myself every time. Like an actor, I felt as if I was living separate lives, with varying qualities in every encounter, with distinct smiles, laughter, likes, and dislikes, each different and yet beautiful in their own way. Overall, it was a surreal and great experience.

So I read, I searched, I lived, I traveled, I grew, I kept to myself, I shed the layers of baggage left by the person I was before. As I kept on my journey, the white fog that emanated with every exhaled breath from my mouth went as the trees began to grow leaves, whose green intensified as the humidity waltzed with the growing heat. So the seasons came and said goodbye. Like a caterpillar, with the seasons, I morphed into someone I had never known, someone to be proud of, someone who was happy.

Then after a year and half of being single, finding happiness, meeting great people, making amazing new friends, reacquainting with some old ones, and starting new businesses, life was good. But then when I least expected it, something changed. I missed the company and intimacy of a woman that came with a relationship.

I started missing the little things we often take for granted in a relationship. I grew tired of conversing with people who still really didn't know me, or the depths of my mind. I missed having someone with whom to wake up in the morning, drinking coffee in our pajamas because the formalities of "looking good" no longer had the same meaning. I missed someone to cuddle in my arms, or caress my hair while we laid on the couch, watching a movie. S**t, I even missed having discussions, because even that required getting to know someone on a deeper level.

I was surrounded by people, yet out of some natural or misplaced feeling I didn't fully understand at the time, I started feeling *lonely*. I missed a different kind of hug, and I missed having an intimate companion who I could call my best friend. Someone who, in all my masculinity, all my testosterone-driven personality, and all my competitive biological nature, intensity, and toughness, could tame me, soothe me. Because while the world saw a tougher man, only she saw a softer side to the man inside, one who belonged to her alone. Someone who one day tells my future kids, "I'll talk to your dad," as she knows she can handle him. I wanted my Clementine Churchill, my Zelda Fitzgerald, my Michelle Obama, my Coretta Scott King, my Zira to Cornelius.

What can beat the act of being intimate and sexual with someone with whom you can be your true self, someone to make decisions with, and call your best friend and family, who once was a complete stranger? Above all, I now know I missed the feeling of love and a sense of nurture. A thought that as a man was hard to express even as I was typing it. But why would emotional expression be any different for a man?

Oppressed, Repressed, Suppressed

Our society has efficaciously *oppressed* a man's ability to express his feelings. Still, men participate in the self-waterboarding act of *repressing* our own emotions. For the sake of serving this very lecture justice and positively affecting our future relationships, the overall societal sexist bias must be effectively *suppressed*.

It's not a martyr competition. By the simple fact of being humans, both sexes face an array of social pressures contributing to the daily stress in our lives. One implied social pressure upon one group does not negate the other; they're simply different. Try not to misinterpret this content as a plea for equality, but a request for reasoning and a bit of latitude.

In the case of a man's inability to express how he feels, despite multiple studies showing no evidence whatsoever suggesting men and women feel any different, the implications of such bring an ample amount of unnecessary complications. These complications include coping with stress for long periods, enhanced signs of frustration, carrying infectious stigmas to possibly good relationships, and on the lighter side of things, sometimes even to the writing of an entire book on the subject and on many other situations to yearning for a nurturing female partner.

In one of many "bartender" conversations (the existence of the term itself is ironic), a man vents the story of how he was dating a lady he liked and, from what he could see after a couple of dates, they had great chemistry and were hitting it off. One day he allowed himself to remove the lid off his feelings and texted her a simple "I miss you," to which her reply was—*"crickets."*

Even I can tell you the limited three times in life I've given into a softer side of me, I have met a 100% failure rate. Thus, why a man cannot give in to sensitivity and risk becoming a "yes man" in a relationship, regardless of how comforting the act may be to him. Arguing for such opposing polarity of acceptance by our female counterparts would be hypocritical at best. The truth is that it would defeat our purpose in natural selection and contradict every point in this book. But allowing a bit of elbow room would go a long way in our relationships.

Stop seeing the occasional expression of stress or feelings in men as weak. Men want to be expressive and tell you they miss you and want you and how darn stressed they are as well. Perhaps vent about the lousy month the business had without fearing not being "alpha" enough. A dominant, providing male also needs support. Hence why after a lifetime of oppression and repression, finding a nurturing, loving partner with whom to share a life, create a family, be open with and real with, to conclusively suppress the vicious cycle discussed here, is so uniquely important to a man.

NINE

What Men Want

THE SEVEN QUALITIES OF AN IDEAL WOMAN

L et me first pose a question to you: *why* do you like the gender that you do? Regardless of your proficiency in science, answering that question from a logical point of view should be *impossible.* Which is the purpose of posing the question in the first place—leaving you with a handful of answers sharing only one possible result: you were born that way.

In other words, you were programmed. Programming that was reinforced in your development. Whether your answer was *I was born this way, God made me this way, biology programmed me*, or even *I don't know*, all answers lead us to the conclusion—it wasn't a choice.

If we can find this to be true, then our thesis should now be easier to interpret. Men are programmed by nature to seek to be loved and nurtured by their partners.

If the term "Mother of Dragons" jogs your memory, then the name "the Unsullied" will too. Both are from the well-known series *Game of Thrones* written by George R.R. Martin and produced by HBO. Like everyone and their mother, I was glued to my couch in front of the TV with some popcorn, ready to watch a new episode every Sunday. If you're unfamiliar with what I'm talking about, the Mother of Dragons refers to one of the story's main characters, Daenerys Targaryen, whose whole purpose is to claim the Iron Throne to the Seven Kingdoms, and the Unsullied refers to her fearless army of highly skilled warrior-eunuchs.

The term "unsullied" itself means *untarnished* or *virginal.* The army of Unsullied were trained as slaves from childhood by different methods, including their castration, so that they would have no desires but to serve the purpose of killing.

In one of the episodes taking place in one of the cities they have conquered, they show an Unsullied soldier walking into a brothel (brothels are quite normal in the show), immediately making the viewer wonder why, if they can't have sex. Moments later, you see the soldier lying down and a sex worker lying down behind him to cuddle him and simply hug him. I found the detail fascinating, because regardless of their inability to contribute to the cycle of life, and the intentions of their slave masters, something embedded in their nature compelled them to seek their most important purpose—their need to be nurtured and loved.

Therefore, like the Unsullied, men in the real world may also be programmed to search and crave the same affection. Yet, to support our claim, we'll go over three critical components in their most simplistic fashion: How the Oedipus Complex affects our relationships; the nurturing biological nature of women and why this is important in a relationship; and how men's societal oppression to suppress their emotions.

The Oedipus Complex in Love

"Oedipus Complex" (the Electra Complex in girls) is a term used in reference to the phallic stage, one of the stages of psychosexual child development coined by one of history's renowned psychologists, Sigmund Freud. The name comes from a story in Greek mythology about a character named Oedipus. Now, understand the story does not explain what the Oedipus Complex is; it just tells us the association with the psychological term.

In Greek mythology, Oedipus was the outcast prince of Thebes, son to King Laius and Queen Jocasta, who prophecy stated, would kill his father out of jealousy to marry his mother when he came of age. Therefore, our unfortunate protagonist Oedipus was raised by King Polybus and Queen Merope of Corinth, who didn't have children. Later, upon finding out about the prophecy, disgusted by the idea, Oedipus flees Corinth, fearful of fulfilling his own predicted destiny and itself. As fate would have it, on his journey, he stumbles upon his biological father, Laius, who he ends up killing due to a small chain of events and a clash of power and egos that led to an argument about who would cross the narrow road ahead first.

Oedipus would continue on his aimless wanders until meeting the gate-guarding Sphinx at Thebes. The Sphinx carries a death-sentence riddle which Oedipus happens to answer correctly and is consequently handed the throne of Thebes, ultimately fulfilling the prophecy after all and marrying his mother, the queen.

Freud used the term to describe the phallic stage of psychosexual child development, which is one of five stages: oral, anal, phallic, latent, and genital. The phallic stage takes place when the child is three to six years of age. Freud argued that during this stage, the child becomes subconsciously aware of his sexuality and becomes competitive with his father from the jealousy of his mother. This biological dilemma will have him compete for the first image of the opposite sex in his life, hence his mother. The idea may sound cringe-worthy to the unlearned ear. It has been the center of controversial debate among scholars, and yet to this day, it holds significant weight in the subject of relationships by professionals in the field.

An article in the *International Journal of Scientific and Research Publications* by Jayson S. Digamon states: "Perceived romantic security in relationships is a crucial factor in determining the psychological health of a person, particularly in healthy functioning of emotions."

Sexuality is an imperative asset to our biological structure, encompassing a lot of aspects of our personality overall. Therefore, it sounds logical that despite being born with our need for sexuality, our anatomical design requires further stimulation through our social and bonding upbringing to fully mature. The same can be said about many other aspects in humans and the animal kingdom as a whole, like our social insight, empathy, love, nurturing, bonding, fear, and more. The majority, if not all, of our encompassing characteristics revolved around the same purpose— socializing. Because as we covered before, socializing is primordial to the experimental erection of consciousness.

In Freud's phallic psychosexual stage described by the Oedipus Complex, the child must experience different phases of the complex, such as attraction, flirtation, rejection, and imitation to develop into a healthy adult who will contribute to life's purpose, which is effectively to create more life.

To better illustrate this, entertain me by imagining little Emma. Emma is five years old and lives in a traditional household with both of her parents. Little Emma loves both of her parents but happens to be quite fond of her dad, as she calls him her "hero," and frequently tells the other kids at school,

"My dad could beat up your dad!" all while mockingly sticking her tongue out. In little Emma's world, her dad is an enormous creature who she's convinced could kick Batman's rear.

As misplaced as little Emma's assumptions of her dad might be, the truth is she doesn't know why if you asked her to explain it in detail. All she knows is that it is real to her, and that's all that matters. When Emma's dad gets home from work, she regularly rushes to beat her mom to greet him and jump to his arms. In school, little Emma draws pictures of her family with her dad by her side for her assignments. An image that now hangs by the fridge.

Little Emma also regularly sits on her dad's lap as she enjoys telling him about her adventures and further adores hearing that he loves her and that she is "daddy's little girl" while he picks her up and kisses her on the forehead. Yet, Emma sees that Daddy kisses Mommy on the lips, to which her only thought is *Yuck!*

Nonetheless, Emma thinks her mom is less approving of her than her daddy, and though she feels terrible for talking back to Mommy, she can't help but feel upset at her and has lately referred to her as "Mom" only. Yet, despite how unfair she thinks mom's treatment is, her dad seems to always side with her mom, which in many ways makes her sad. When Emma's parents notice she's a bit quieter than usual, playing in her room by herself, both of her parents come in to remind her how much they love her, and little Emma's sorrow eventually subsides. *I guess it's not that bad after all*, little Emma thinks. In her kindergarten class's last drawing assignment, she draws her family again but with her in the middle this time and a big ol' smile on everyone's faces. She tells her teacher how she loves both mommy and daddy equally and requests the drawing on the fridge is updated with this one.

What little Emma doesn't understand is that she subconsciously learned a fundamental lesson in her life from external stimuli that registered with her despite her awareness of *it*. Little Emma has no idea she needed to face the subtle rejection of her dad to understand that he indeed belonged to Mommy as little Emma doesn't know of her heterosexuality, or her sexuality at all, for that matter. An effect that would later make her imitate Mommy instead of competing against her.

The Oedipus complex explains that little Emma was learning to "flirt" with the opposite sex as she was attracted to the presence and strength of her daddy. In this preferable and fictional world, Emma will effectively have a higher chance of finding an ideal partner with similar characteristics of her "hero" and maintain a healthy relationship by implementing the good traits of her mother.

In the same context, a male will learn to seek the natural nurturing traits of his mother while implementing the qualities of his father. In males, the clash between testosterone-filled egos can last up until pubescence and appropriately conclude when the alpha male hierarchy and dominance gets established by the head of the household. In lions, this clash is quite literal, and there can only be one victor of the herd, which happens to be the older male lion in the majority of cases, effectively kicking the adolescent male out into the world.

After grasping the concept of the Oedipus Complex, it becomes easier to see how detrimental to someone's love life and the relationship it can be to have an absent, disrupted, or interrupted childhood development process. The possibilities of these unfortunate, hurtful, and non-ideal situations are practically endless, from divorce, absent or inattentive parent(s), sexual abuse, or simply bad parenting. Yet, the effects of such misfortunes can be seen through the entirety of an adult's failed, toxic, or vicious relationships if one fails to engage in introspection or seek proper therapy. The person suffering such consequences has to realize this themselves and want to help themselves to proactively work towards a healthy and happy future in the life of love.

Regardless of the contentions against Freud's psychological theories, the art of knowledge permits the questions that bear a discussion of the topic at hand. Many contemporaries of psychology today continue to support and debate the subject. For example, Jordan Peterson lectures on the desire for redemption and the proposed theory that women are deeply attracted to the negative characteristics of our primary caretakers. The unconscious feeling of not being able to live without them is more like wanting to go to the past and fix the relationship with our parents.

A Woman's Nurturing Nature

Historically, women are the caretakers of humanity, the mothers of the world, the bonding agent to child-rearing. It's no surprise that when we look at a woman's biology, we see that generally, women tend to be better nurturers. Their nurturing nature is implied by merely looking at the higher levels of oxytocin (the love hormone) a woman experiences through pregnancy.

Coincidentally, from a factual or personal, objective or subjective, external or internal point of view, men both express and tend to gravitate towards a more affectionate woman who happens to be more in touch with her nurturing side.

After waking up the first dreadful morning following a breakup the night prior, I hung out with one of my best friends, Sketch. While having an in-depth conversation about life and relationships and joking about adventures with the ladies over some drinks, he had this sincere moment of vulnerability and admitted something I sincerely appreciated. A bit more serene in his demeanor, he looked down, lowered his voice, and said, *"You know man, I'll keep it one hundred percent with you. The truth is that despite all these stories, not having a girlfriend can get lonely. That's why I bought my dogs so that they can keep me company in those times."* Few times has an expression been more meaningful to me because, at that moment, a great friend of mine let me know I wasn't the only one feeling the need for intimacy and affection that comes from a relationship with a nurturing woman.

But aside from personal tales about men's need for affection and nurturing, science may be able to broaden the scope for us from this perspective. Upon listening to another talk by the much-respected clinical psychologist Jordan Peterson, he made the point that women tend to be attracted to bigger men, who have broad shoulders and a good upper-body frame-to-waist ratio. The reason for this attraction is due to a woman's biological need to protect her offspring during the vulnerable months of pregnancy and the first nine months of an infant's life that requires a mother's full attention.

Likewise, a man's search for a nurturing companion might not be for himself. Still, perhaps in conjunction with the point mentioned in the section above, a man also subconsciously longs for the affection and nurturing of his partner to gauge whether she will be a good mother. Just like the physical appearance of a woman attracts a man sexually is indeed a genetic code, the male brain translates into a healthy, fit womb to nourish a fetus for nine months. But how else do men choose the women in our lives? How else do we decide on the woman with who we want to spend the rest of our lives?

The Ideal Woman

If you thought sex and nurturing was all it required to get a quality man, not so fast. We, too, have characteristics we look for in the woman with who we want to settle. Because as I've said before, we want our next partner to be our last.

On the other hand, if you date a guy who tells you otherwise, why else would he be dating you? And a better question still is, why would you continue to date him? I advise you not to waste your time, since this is one of the prime characteristics of a man-child. Whether he knows it or not, this guy is not ready to be in a relationship, and therefore shouldn't be in one. Not to worry; make it easy for him and walk away. While it's perfectly normal to be uncertain of jumping into a relationship right away, this is a decision that should not get past the first two months. Otherwise, he's just in it for the sex.

After nearing the end of the seventh chapter in this book, soaking up all the knowledge, and hopefully putting it into practice, you deserve a quality man. But what does a quality man look for in a woman?

Personally, I look for four qualities:

1. Health and beauty (which is a combination of nice facial features and a nice body, so preferably someone into fitness)

2. Someone goal-oriented and who has a competitive drive is paramount for me since I'm very driven and goal-oriented myself

3. Intelligence, since it's an essential quality in decision-making and tends to correlate with knowledge

4. A sense of artistry, since I believe it gives me a good understanding of their emotional spectrum

Being the complex animals humans are, beauty is in the eye of the beholder, and we all have different preferences and tastes that vary from person to person. Nonetheless, after asking every single one of my male friends, engaging in self-discovery, asking relationship coaches, and gathering what over twenty-five respectable articles had to say, I've compiled a list of the top seven qualities men look for in women.

The Seven

As you read on, keep in mind one word—middle. As we go through the qualities good men seek in their potential forever mates, notice they follow the paradoxical principle of life, yin-yang, life and death, male and female, good and evil, darkness and light, failure and success. Life is about finding balance, and the qualities we seek in our ideal partners are no different.

1. **Physical Beauty**   2. **Intelligence**
3. **Ambition**   4. **A Kind Heart**
5. **Sense of Humor**   6. **Femininity**
                              7. **Confidence**

1. Physical Beauty

The first thing we—both men and women—notice about each other is physical appearance. While we all like aesthetic facial features, there's not much we can do in that department if we weren't gifted with natural beauty. I pluck my eyebrows and use a facial scrub and facial lotion. Still, despite sanding my face ten times a day or moisturizing it by diving into a bath of lotion, I won't be looking like Henry Cavill any time soon. It's always a great habit to take care of your skin, and while I personally like ladies who touch up with just a bit of makeup, how much or how little you want to wear is obviously your personal preference.

The good news is that our head is only 10% of our overall body mass, which means we can work plenty on the other 90% and look hot and spicy for our next life partner. Though I don't need a biological anthropologist to tell you the first things a man notices about a woman are her butt and breasts, scientific publications do support this claim. As we covered before, this is due to a man's wiring looking for a healthy mate who is most suitable for producing life. Meanwhile, no (heterosexual) man can tell you why they are attracted to those features aside from telling you we bite our fists when we see them.

You should already have a good idea of your preferred workout from our breakups chapter. But if you'd like more ideas and details, make sure to check out my blog on my website where I go into specific detail on health and exercises you can practice getting into the best shape of your life.

2. Intelligence

Scientists argue whether intelligence is something we are born with or something that can be improved over time. We're indeed born with natural capabilities and differences in IQ levels that are notable from the early stages of child development. Whether that's due to the number of neurons in our frontal lobes, the actual fluidity of electrochemical communication between the different areas of the brain or something else is still a subject we're learning more about frequently. What's also heavily debated (in science, what isn't?) is whether intelligence can be improved or not and while most evidence points towards a solid and disappointing no, that only applies to something known as fluid intelligence. What *can* be improved is **crystallized intelligence**.

Crystallized intelligence is the accumulation of knowledge, facts, and skills that you acquire through life. As a man who considers himself adequately intelligent, finding a partner with a similar intelligence level is an essential factor I look for in a woman. Trust that any "quality man" out there waiting for you to catch him looks for an intellectually competent and like mate.

While I'm not looking for a Katherine Johnson or a Hedy Lamarr since I'm a genius myself, I would appreciate a smart lady with logical, good decision-making skills who can engage with me in conversation about some of the wonders of the universe. Therefore, if you're ready to catch that quality, smart, handsome-devil, son of a b****, you're already ahead of the game by reading this book. Reading novels can be amusingly addicting, but picking up an informative self-help book never hurts.

3. Ambition

A quality man has his finances in order, takes his career path seriously, and is always looking to increase his net worth. It's only logical that such a man looks for his female counterpart to do the same to an extent. Trust me, ladies, there are few things as sexy as a woman who can talk to you about her goals in life with confidence and determination (my eyes rolled back in my head as I typed that). Despite what some ladies may believe, a man's desires can pretty much be summed up as someone who wants a best friend for life. As you learned in our previous chapter, happiness is a state of own. In the same regard, it's exciting to meet a woman whose life I can contribute to and vice versa rather than someone whose life I have to support.

Again, our previous breakup chapter fits like a glove in this section, since we've already put into practice jotting down some goal-setting. Discussing paths of success and career paths is opening a whole other can of worms one can get into on a different occasion. Instead, try to answer this question: *Where would I like to be in the next five to ten years?* Once you answer that, great. Now, draw out the map to get there. Remember, from point A to point B. Piece of cake, you got this!

4. A Kind Heart

This trait may go without saying, but it's still an essential factor a man looks for in the ideal lady. A kind heart means that a woman is capable of empathy, a quality that goes hand in hand with someone supportive, understanding, and kind. Besides, it's the umbrella over a woman's nurturing sense. By now, you might have an idea about my grievances regarding a male's oppressed emotional expression by society. So coming home from a world that offers no more guidance or empathy than *fix it* or *suck it up* to meet a lovely woman with whom to express freely, being completely yourself, while enjoying a tasty carbonara and sipping (perhaps gulping) a glass of chilled sauvignon blanc—what else needs to be said? That's the stuff of life right there!

You, too, face different societal pressures. You, too, have faced your share of jerks and incompetent men. Let's not let our memories make decisions for our present and, in consequence, ruin the possibilities of you finding good men. Don't allow invisible memories of unworthy men to bully you not to embrace your softer side. I assure you, there's a real man who is worthy, waiting to snatch you away.

5. Sense of Humor

"Someone you can have fun with, and that gets your sense of humor. I feel like I married my best friend, so it's good to be able to have fun and joke around." Those are the words of my friend Elliott Gular talking about his wife as I queried the minds of my many male friends.

A sense of humor is a two-way street, an element from which chemistry is made. What's an anecdote, sadness, stress, a breakup, if we can't laugh about it down the road? Laughter is the universal cure to stress and the key to love. As you go on your quest for your ideal man, make sure he is someone who knows how to let loose at the right moments, embrace the kid inside him, and joke about the intricacies of daily life. Someone who, after years of a healthy and long-lasting relationship, you can talk s*** to, prank, make mistakes with, and embark on a life where you know that no matter how tough it'll get, he'll be waiting to console you, support you, and make you laugh, as I assure you he's looking for the same right now.

6. Femininity

Being a man or a woman does not automatically translate into being masculine or feminine, just like nothing prevents a man from being feminine or a woman masculine. They are aspects of our genders to be embraced. Masculinity is usually associated with characteristics of dominance, strength, and power that ultimately give off an aura of protection and leadership. If you're not fond of insects and you see a spider roaming around in the bedroom, you'd look for me to take care of it (a moment in which I'm not fond of masculinity). At that moment, it's safe to say you'd feel protected and supported (yes, killing a spider is a big deal). It doesn't mean you expect me to be a protein-craving, macho, John Wick, tough Mafioso twenty-four-seven. Or at least I hope not, because I can tell you right now it's not on my agenda.

Similarly, there's nothing wrong with an independent, self-sufficient woman who can handle herself. Trust me when I say there's nothing sexier than or as arousing as a woman who knows how and when to embrace her femininity. A man wants to feel like a man, and nothing amplifies that masculine feeling as its polar counterpart—femininity.

But what does being feminine mean? The truth is no one definition can define femininity for you. It's something you must describe yourself by embracing your sexuality. It's not even exclusive to gender, and feminine traits in men can often be perceived as highly attractive, as Robert Greene tells us in his book *Seduction* (2001). Breaking the chains and tossing away all of the world's "status quo" imposed pressures, what does being a woman feel like to you? Forget about the man on this page. There's no one here but you. Does it mean wearing clothes that complement your figure? Is it embracing your softer or higher pitched voice? Is it embracing your sexuality in the bedroom?

As you move on to future chapters, you'll learn that in the art of communication, for example, two rights (truths) can co-exist. One does not have to negate the other. While you can be a dominant, aggressive businesswoman at the office, you can also embrace your natural feminine prowess. As a matter of fact, "The Breaker of Chains," a.k.a. Daenerys Targaryen from *Game of Thrones* was a leader with thousands of men under her command, conquering cities as she moved toward her destiny, but I can guarantee you *no man* was saying, "yuck." Embrace the woman inside and give the man of your future what femininity means to you.

7. Confidence

Not arrogant, not cocky, but confident. Arrogance is thinking we're more than what we are, and being cocky is compensating for what we lack (hence the word.) On the contrary, being confident is someone who has found that sweet middle spot and knows their worth. Attaining a sense of self-worth is having a real sense of who you are as a person and understanding why. Be it your achievements, craft, profession, your body, your financial status, or circle of friends. It is an understanding of the work you've put in to obtain whatever you consider contributes to your sense of worth. It's a perspective that can only be attached to a sense of humility, as confidence can only be obtained and never given.

Men look for a woman who stands on her own and is *truly* independent as f***, which can only come from someone who practices introspection and has found happiness within herself.

Confidence is not bound to being right, knowing it all, or always being a leader, or being independent. It's as much about knowing when to lead as it is about knowing when to take a step back. It is about confidently knowing when we are sure of an answer as it is about saying, "I don't know."

Every good man craves a confident woman who knows when to use her strength, wisdom, or determination. We die for a woman who has the confidence to be different roles and is unafraid of stepping over the lines of the imposed historical boundaries put on women from fearful men, one who knows when to speak and when to allow the loudness of the silence to grow. A woman confident to be a shoulder to cry on in our moments of weakness and challenge our growth. I believe we live in a different era where we are redefining our gender roles every day. Further, I think a modern-thinking man recognizes a woman's gifts and embraces differences that require an ideal counterpart—a confident woman.

PART THREE

The Stages of Love

My study _ _ _ _ _ _ _ _ _ _ _ _ _ _ _ _ _ _ _ _
So know that my birthday happens on May
and the day is the ninth.
Which explains the lack of apology, and
also how a person is socially kind.
Yet it' s as simple as doing the math,
to finding the word to this simile-dwell-blog,
Concluding the answer to an empathic aftermath,
The answer is ....?

TEN

Dating

THE WORLD OF LOVE

his chapter can be summed up by saying: ladies, control the sex. There, it's settled, on to the next chapter.

Men will hate me for saying this, but if you listen to anything, listen to this. In the dating stages of love, *control* the sex. Why? Well, as you now know from having learned some of men's dirty little secrets, sex is a man's only original intention, and thus the key to them falling in love. Therefore, there are two approaches to this: one, either *do not* have sex with your dates during the first three encounters while continuing to flirtatiously flaunt all of your aesthetic physical prowess in their face. Or two, if for the few and special reasons (real chemistry) you end up sleeping with him on the first date, make sure you're a sex goddess, or else you'll be wondering why he happens only to want to hang out on Friday nights.

While women might like good sex, men go crazy for good sex. If you follow this simple game plan, they won't know what hit them. On the contrary, a word of caution, precisely because you enjoy sex as well, women sometimes fall into a self-laid trap of "sex without strings attached"—an utter fallacy. If you happen to be thinking in this regard, do it fully aware of what you're getting yourself into, because in the likely case this stems from any type of insecurity or search for closure, this will only hurt you down the road, as I assure you. It is not a game you'll be able to win against a male. This agreement is the ultimate fantasy contract offer for a male, and you can be sure he'll finish signing before you can say "no strings at—" And if that wasn't enough, be ready to renew this contract as indefinitely as possible.

As we have touched upon previously, it just doesn't seem women need physical options to the extent of the materialistic preference that men do, even though women are also attracted to physical appearances. As Jordan Peterson has expressed, women generally look for a provider.

Data based on historical records and ethnographic interviews pieced together as life histories by anthropologists indicate that in forager societies, men who can put more food on the table and create wealth are most likely to have more partners throughout their lifetime. As we emphasized with men and sex, despite sex always being a man's intention, it still shouldn't be their *sole* intention. The same applies to you in this scenario. Just because of data, biology, and whatever other mumbo-jumbo points to the overall generalization that women look for a provider. Equally, a man who is purely a provider shouldn't be your sole intention or the only quality for which you should look. From friends to acquaintances, or during open discussions, I regularly see women fall to this grave notion of dating the ultimate provider who is *nothing but* a provider!

Therefore, be brave enough to pass judgment on yourself and take the blindfold off this wishful illusion by bearing in mind that if a man's worth is as much as the weight of his wallet, then what is he buying? I hate to say it as much as you hate to hear it, but if you choose men strictly for their financial appeal, then I hope you're willing to accept they're not with you only for your charm.

On the other side of things, if you're ready to start dating, you should take a moment to see how far you've come and how incredibly proud you should be of yourself. Because unlike a lot of people, you've beaten the status quo and set yourself on a promising path to finding true love and a healthy relationship.

The World of love

This chapter intends to teach you how to a contender in the world of love. Most people don't know what that world implies. As you might know, it means playing a dating game, despite us wanting to. But you've been a part of it regardless of being aware of it. We often get involved in this game unknowingly from sheer insecurity or the fear of the vulnerability imposed by the thought of falling in love.

It is this fear of vulnerability-turned-insecurity that continually nourishes the never-ending cycle that we refer to as the world of love. In other words, we compel one another to play this often-senseless game. If you can visualize fear as a physical object, all we're doing is playing hot potato with it, throwing it around from one to another. This bizarre social conditioning is best known in psychology as classical conditioning.

Does the name Ivan Pavlov ring a bell? Classical conditioning is learned actions or programmed reactions from associating two paired stimuli, for which later the process is provoked by only one of the incentives. For example, in the famous Pavlov's dog experiment, he rang a bell before serving a dog its food multiple times, teaching the dog to salivate upon seeing the food. Ultimately, the dog's learned expectation of food comes at the sound of the bell, causing it to salivate with just the first stimulus, regardless of getting food. Similarly, we have learned that our unfiltered emotions are commonly met with a sense of unappreciation, callousness, and an advantageous demeanor from experiences with past partners, teaching us to keep them bottled inside.

Love Vs. Being in Love

I believe people would benefit from ridding themselves of the notion of "being in love." Being in love means being drunk on your emotions, feeling that person's feelings are more important than your own, and, frankly, being in an illogical state of mind. Even in science, the idiom "love is blind," as you may recall from Chapter One, comes from the fact that the system responsible for making critical evaluations of those with whom we are romantically involved lowers its activity.

When things are working out, which is usually at the beginning of the relationship during the honeymoon stage, being in love feels great. Nevertheless, letting those emotions run rampant is also allowing them to rule our minds, instead of allowing our minds preside over them; like a conductor to an orchestra. Even still, no one is suggesting to stop feeling, but rather to stop making decisions for your romantic life purely from how you feel. Because as you may recall, we are a potent cocktail of chemical reactions that can only be gauged by consciously practicing different methods of engaging intellect, proactive thinking, and more. Allowing our emotions to influence us so broadly is undoubtedly to let ourselves *fall in love*, and when *in love*, indeed—*we fall.*

People naturally become addicted to their emotions. After all, what do you think most drugs are if not little soldiers inhibiting neurotransmitters or blocking them, causing us to get high on our own supply?

Being in love will make you say things like, *"you're my life," "you complete me,"* or *"I can't live without you."* But you don't complete me, because I *am* complete; you are not my life because I define what my life is since it is mine; and though I sure don't *want* to be without you, I can unquestionably live without you, and the same should apply to you and everyone else.

Your value should depend solely on you; it belongs to no one else. Your self-esteem is the esteem you hold for yourself. I shouldn't be your life, nor do I want such a burden. Asking that of anyone is alleviating our responsibility for the person inside and is plainly and outright selfish.

Ideally, I want us to share our happiness, not make each other happy, because if **I** haven't found out how to be happy or what makes me happy, how could I ever expect **you** (a newcomer) to know what makes me happy? Instead, I want to contribute to your life, to add my value to yours, to strengthen the person you already are, and **vice versa**. Then, with time, as we grow, and our chemistry broadens, and our *respect* strengthens, the admiration for one another that stems from that will inspire acts of kindness. But there will be moments when we don't feel so kind; moments we can work together and learn to *compromise*. Above all else, it is allowing the culmination of all elements acting in unison that supersede the pronunciation of love.

We forget an essential truth from our ancestors. The original purpose of words was to represent actions, not replace them. Language came after merely to describe them. But it seems this is alien to us now. We forget the power of *showing* our significant other we understand their love language.

Therefore, if we practice our actions more, demonstrate our love, support, and care to each other, then in the moments we say "I love you," it would be merely a *reminder* of what we already know, not serve as *evidence* of our love. To love someone is a process, something that grows like a living organism. You can hinder its growth, or you can feed it, water it, and nurture it, but not *fall* into it.

Dating

If you're ready to get back into the dating scene, it's crucial for you to learn the rules of dating and how to win at it to achieve your desired goal—a loving and healthy relationship. Do you remember the movie Days of Thunder (1990) with Tom Cruise? It is a great driver—meets the rules of racing. Similarly, all the work and growth you've established through a step-by-step process in seven chapters would have been all in vain if you don't know the world in which you're involved.

To make it easier, I've broken down the steps into a neat little dating starter pack with three fundamental principles that together house a handful of insight, advice, tips, and tricks with philosophy as a foundation. *If I've put so much emphasis on finding and improving myself, why must I still play this game?*

Coincidentally as I was writing this chapter, I logged into one of my social media accounts, and the first post I encountered was relatable to our topic and the overall essence of the book. The post read, "*Everyone you meet comes with baggage. Find someone who loves you enough to help you unpack.*" As I read that, I couldn't help but think, *hmm, I wonder how someone finds someone who just loves them with all their baggage?* I extend the question to you—how? This mentality was one of the primary motivations to write this book. I wonder, is there a bank? A line? A service I should be aware of where partners are prepackaged to love me? Though well-intended, the quote is a "accept me as I am," lazy, victim mentality that can lead to dangerous misconceptions. I'm afraid reality is not so gullible.

You can't *find* someone who simply loves you because we *influence* people to love us. It's not something that falls from the ether into our laps. It is something we manifest into existence, something we will into our lives. It is something embedded into every one of us, regardless of reading this text or being conscientious or not. Otherwise, why do we have a style of clothing? Why get a haircut if not for that subconscious purpose? And if we are already doing all that subconsciously, why not improve at obtaining what we want morally? Becoming aware of all that love requires is what separates the chosen from the ones who *choose* their partners. You'll have to decide which one you are.

There's a professional poker player named Daniel Negreanu. When reading his book *Power Hold'em Strategy,* he mentioned an interesting theory. He noted you never put in your chips against an "all in" rookie player because they overvalue their position, and if you bet against them, they may get lucky and take your money. In the same context, you are the wise and educated player in this scenario, entering the dating phase in the world of love. Therefore, moving forward, you'll be wise to remember the underlying principle to the overall dating philosophy we're discussing—this is an **act, seduction**, and **manipulation**.

It's important to understand these words so that they're not taken out of context and apply them correctly to our dating strategy. A small concept I use in business and sales will help you differentiate between *win-win* situations and *win-lose* or *win-lose-lose* situations. In that light, you'll learn that *acting* does not imply being deceitful, *seducing* does not mean brainwashing, and *manipulating* does not mean domination, totalitarianism, or being commanding.

Win-Win Vs. Win-Lose

Let me introduce you to the surface of the business world. In sales, unfortunately, the best salesperson and the one often celebrated by the company is the most deceitful, lying, unethical person you can imagine. I always happened to oppose the corporate definition of what a good salesperson is, hence why I quit. Business and sales are an art that need not be disrespected by thieves who excuse calling a robbery a sale.

The best and most trustworthy businessperson is one who uses words as his mastery, one who persuades a customer into whatever product they're trying to sell because they genuinely believe in that product. Not one who spits lies, empty promises, coerces, and creates a negative experience for the customer. A deceptive salesperson creates *win-lose-lose* situations—win for *them*, a loss for the *company*, and a loss for the *customer*.

If someone's ever sold you something that made you feel swindled or you called to cancel or return the product as soon as you got home, you have been a victim of *deception* and have a good idea of to what I'm referring. Conversely, if you get home and feel proud of your decision, perhaps even want to tell someone about it, you've been part of *inception*.

Likewise, a womanizer (or a...manizer?) has a plan that will create immediate results for themselves—such as sex for men, or gifts and expensive outings for women—that have an immediate gratification (dopamine fix—reward system) for them, yet a negative impact on the other person. Both in the business and love world, these individuals are people users; they are dopamine addicts who gain their fixes using the collective instead of creating *win-win*, meaningful experiences for the long run—win for them and win for you.

Neither of the two former outcomes is the intention of this text. I am *not* interested in using women to fulfill my sexual cravings. I want sex, but not at the expense of hurting or demoralizing people. Likewise, men are *not* for your entertainment, and if they are, what does that say about you? Are human wallets and human toys all our value affords us? With this approach, you'll keep bumping into the unappreciative males we've been discussing.

We are in the business of *win-win* situations. We are here to learn all the secrets, the map, the rules, tips, tricks, turns, and curves to gain all the tools necessary to face the world of love out there. Whether you use those tools for a useful purpose or a bad one is on you. In this text, we are in the business of siding with the middle.

It is crucial to continue driving this point home until it's clear and engraved in our minds because, as with any tool, weapon, or knowledge, it can be used for wrong reason by ill intentions or by merely misinterpreting the context.

In the business world, an ideal businessperson is one who is willing to *bend* the rules for a positive outcome, not break them, *use* the furthest extent of the law, not break it. An ethical salesperson manages to sell the product without inflicting a negative outcome on the company or a customer by omitting the truth, not by lying.

For example, I sell you a fashionable piece of ladies' clothing by telling you it's good quality (because it is) and giving you a fair price on it, but I omit that I bought it for 20% of the price I sold it to you. Telling you my buying price could deter you from buying it. Otherwise, in the end, you still have a piece of quality clothing you liked, and at a fair price, while I still made money. Furthermore, a good salesperson is the one who sells without the implication of selling. They call it "building value," meaning they'll talk about the value of the product, the value it adds to your life, and the value of its practicality, not the cost of the product, the price vs. the competitor, or the cost it will subtract from your wallet.

The customer knows what they're willing to pay for any given item; there's no need for me to remind them of it. All the same, your date knows what they're investing in, and hopefully, you do too. There's no need for us to be forwardly outspoken about each other's intentions. In the world of dating, you are buying—procuring *love*. Therefore, there's no need to hurt, use people, and ruin lives. We should do acts that induce something positive in both parties.

Acting does not imply being deceitful. The definition of acting is being adapted or designed or serving temporarily. Whereas being deceitful is being false and misleading. See acting as the temporary magnifying of our natural traits, where deceit is lying about our characteristics or straight-up making them up.

For example, I've always gotten three main compliments: my intellect, looks, and being a good dude. So, I'll try to impress my first date by saying something intellectual or use my appearance to be more flirtatious with *Zoolander*-esque mannerisms (irresistible). And knowing that nice guys finish last, I'll simply employ some qualities of being a "good dude," such as being a good listener and perform occasional chivalrous acts with confidence. On the contrary, I'm not a big fan of camping (glamping's okay), therefore saying I am, while offering to go on a trip for the weekend for no other purpose than getting into my date's pants because I know *she* does, is deceptive.

It is the same with you, so now it's your turn! Write down three of your favorite qualities that will lure, and trap the quality man on your date:

1. _____   2. _____

3. _____

Seduction, similarly, does not imply brainwashing or indoctrinating our date. It means creating a two-sided benefit and transaction suitable for both by using our prowess to fulfill it. "You want sex, and you're not going to get it until I have your full attention." It is simply employing a more realistic approach to gaining your goals.

As a writer, I had to seduce you to grab your attention in the first paragraph, page, and chapter, to get you interested in reading more of what I have to say. Otherwise, if I would have started ranting about how misinformed people are, chances are you would not be on this page right now. So, are you hurt? Did I lie to you or brainwash you? Hopefully, you've gained something new from continuing reading.

In the same context, you are merely employing a more realistic approach in knowing what your date wants, what you want, and being the leader in managing that exchange. It's about the cute outfit you'll wear, the perfume you'll have on; it's the sensuality and sexuality that are not the means to an end, but a means to much more—the conduit to love. It is embodying all of your goddess prowess to all that he wants, and that he just won't get unless he earns it. He must first listen and learn about you and show you genuine interest—not only pursue your sexuality. But more importantly, not until you feel content because he fulfills the qualities of what you look for in your life.

Manipulation does not imply domination, totalitarianism, or being commanding. As we've stated, it is not to employ bullying tactics, but to persuade our date. It's like the person who jokes about someone for the satisfaction of making people laugh, except the only person not laughing is who the joke is toward. That's the same reason I never liked late-night talk shows since their entire talent revolves around making fun of their guest and diminishing the artists' work to increase their ratings by methods of cheap entertainment. I've always been a big fan of Conan O'Brien because he uses himself as the guinea pig for comedy.

In dating, it's no different. We can certainly use some bullying tactics to draw insecurities out of our date, make them uncomfortable in our "superior" light, and ultimately mind-f**k them, but how is that beneficial to you?

So, I've included some quick points of fatal attraction that will give you an edge on the game and put you miles closer to the finish line.

Appearance

Our appearance says a lot about ourselves and our personality, from what we like, to the lifestyle we might have, to our likes and dislikes, and to an extent, even our intentions. There are few things as attractive in a man's eyes than a woman who owns her sexuality.

Relating to attire, this is someone who knows her qualities and displays them proudly, yet she does so without being the center of attention. Someone whose appearance lets me know she understands my manly desires but will not come unwarranted. It's a show, not an offering; it's a runway, not the red-light district; it's the hint of cleavage, not the sisters on display; the short top, not the crop-top; the trace of makeup, not the "who's underneath?" The same should be imposed on men—it is the fitted jeans, not the painted-on jeans; it's the watch, and fashionable attire, not the diamonds, earrings, and everything-but-the-kitchen-sink look.

Not Looking to be Entertained

This is your moment to shine and reel him into who you are. Few things can shift a man's focus from sex to wanting to spend time with her like a woman's intellect. Few things are as seductively attractive than a lady who can have a good conversation. Someone who gives open answers, asks questions back, and has a sense of humor, letting me know she's not expecting a clown to entertain her. That's marriage stuff right there; it's captivating, it shows confidence and intelligence, and it flirts with a man's curiosity to want to know more about her.

Embrace Femininity

As you read in the previous chapter, men are suckers for a woman who can be feminine. As you learned, no one can tell you how to be feminine—that's a state that belongs to you. What I can tell you is a man goes crazy for a flirtatious woman (lip biting), who giggles, plays with her hair, maintains eye contact, and subtly checks me out, telling me she's unafraid to show me she likes me.

Social Views

Now I hope this is not the part where you burn this book. We all have social beliefs, creeds, political affiliations, and a personal stance on all of them. Leave that at home, there's a time and a place to have those discussions, and your first date is hardly one of them. The first date is about the chemistry, the laughs, having a good time, getting to know each other, and finding out if that's the right person for you. Especially if it is a subject you know could potentially upset you if you meet an opposing view.

Getting to know someone is about being respectful of their opinions out of the simple knowledge they are as entitled to their views as you are to yours. If your perspective on a specific topic is exclusive, then make sure that's something implied lightyears before going on a date. Perhaps let them know as you're messaging each other or possibly even consider writing it on your online profile.

It's not my recommendation to bring such a mentality when looking for a date, as I find it a bit obtuse, but hey, to each their own. That's what respect is about—I respect your opinion and vice versa. It is *not* about imposing our view on someone else.

A Gentleman Meets a Lady

Remember our talk about chivalry a couple of chapters ago? Performing actions without a predetermined motive or notion by understanding the competence of women while not undermining the intentions of men. There aren't many sights a man finds more attractive than a date who gets up to use the ladies' room and comes back with a round of drinks. It's just as lovely when she offers to pay, but lets us pay after all. Trust me when I tell you, this is a major turn-on for men!

It immediately raises a woman's value in our eyes. It communicates that she likes us, that she's independent, and what's more, that she's choosing to spend time with me for more than the immediate benefit—stakes raised.

Smells Good

If we are potentially going to be intimate at some point, shouldn't we communicate something about our hygiene? What else could possess such polarity of being a complete turn-on or turn-off than our hygiene? And what could be a better way to communicate that than smelling good? It's an even playing field for men and women in this department.

We shouldn't smell like we bathed in cologne or perfume, but perhaps wearing a bit more than our natural aroma would be pleasant too. I don't think most men mind if a lady is not wearing perfume at all as long as her scent is one that transmits good hygiene. A good rule of thumb is if people notice when you walk by, perhaps your scent is too strong. Our fragrance should compliment us, not attract attention.

What's after your date?

So how did your date go? Did you like him? Would you like to get to know him more? Should you now patiently wait for him to contact you? But about if it's been a day or two, and he hasn't asked about a second date? Here's how to know if he's really interested or just playing "the game" for sexual gain. *Bring the game to him and ask him out yourself.* It will prove interest, confidence, and will get you to your answer faster. When you ask a man out, though, do it from a position of *interest* but not *need.* Something like:

"Hey, so I'm going to the Natural Science Museum tomorrow. Would you like to join me? Let me know."

Instead of saying you'd like to do something *together,* or with *him,* which communicates exclusivity and tells him he's the focus of the trip, asking if he'd like to join you implies the activity exists with or without him. Saying "let me know" implies there's a time limit on his response, which if he allows expiring, he misses out on hanging out with your cool-self. But it also lets him doubt if you'll be going with someone else. Now the decision is on him, and so is the pressure

Now whether you take the initiative of inviting him on the next date or he contacts you, make sure to switch the scenery to an activity during the day and one that doesn't involve sitting down to drink right off the bat. Suggest a museum, a nice park, the movie theater, or whatever you like that will shift his programmed focus from *drinks + night* → *possibility of sex.*

Shifting the scenery will send him a message that this is a woman he's dealing with; and will effectively separate an average fella from a quality-man.

Crossing Cultural Boundaries

As forward and evident as it might seem to many to say there's nothing wrong with having cultural preferences, it sometimes might not be as apparent to different cultural groups. I proceed with a self-imposed word of caution since ethnicity is a sensitive subject in our day and age, yet one I find essential to address.

As we proceed with the text, as always, remember to do so with an open mind, and with the understanding that everything you read within the boundaries of this book has no leaning on social issues, let alone following a political agenda. What it does attempt is deciphering what the world of love is—regardless of the steepness or difficulty of a subject.

Whether you're white, Black, African American, Latino, Hispanic, Asian, European, Indian, Native American, or a combination of the above, what's certain is you are, above all else, human. Yet we sometimes face pressures from our cultural group or immediate family, and dating outside those cultural norms might be frowned upon.

Therefore, what if you have a personal preference for someone from another culture? There's absolutely nothing wrong with cultural preferences. We all have preferences. Don't ever let anyone guilt you or bully you into a title. You are not discriminatory, racist, or bigoted, just because you have preferences. What you have is the vision and the courage to accept who you are and what you want. If you can relate, you might have found yourself wondering how to date someone of a different culture at one point or another. On the other hand, if you can't relate to this section, or you find it offensive, that's not the intention, by all means, please proceed to the next section.

What to do about the pressures imposed by dogma? In the general context of preferences, you may like the color red, and I happen to like the color blue. Yet, you don't see us explaining why we have favorite colors to each other, do you? Being that color and culture, when applied to people, can be used to discriminate or insult, it's understandable why our society might be sensitive to the overall subject. But, as you'll learn ahead in our Communication chapter, two "rights" or "truths" can coexist. In this scenario, it's three. It is true we should be careful of how and when we voice our preferences if we do at all, as it's also true that we have preferences and you shouldn't have to deny them, nor avoid the subject. At the same time, it's also true that having an inclination does not mean that you are biased.

For example, I, as you may know from reading the previous chapters, I tend to look for four main traits in a woman. That aside, I have a thing for white women with blonde hair or raven-black hair and pale skin, resembling the Kate Beckinsales of the world (I melt). But that doesn't mean it overshadows the other four traits, nor does it mean I don't love Latinas, Black women, Asian women, or anyone else. My mom is Mexican, my dad is Black, I have cross-culturally dated all of the above, I have a German first name, a British last name, a previous Spanish family name (what's that about?), and I happen to have light skin and look nothing like anyone in my beautiful family, (perhaps I'm adopted and my mother hasn't told me).

What I'm trying to tell you is that social pressures exist, and if the same example applies to you, it's completely fine. There's nothing wrong with liking someone who looks a bit differently than you—you're both human, and love has *no* boundaries.

The first step to take for love to cross-cultural boundaries is to admit it (rebel out). The second one is to speak about it, be unafraid to be outspoken about it. Need an excuse? Yeah, I'm human and like the human race indiscriminately—period! Some of our families have a "don't take one of our good ones" mentality, which can be somewhat uncomfortable. Not to mention, ethnic identification, which from personal experience, friends' tales, and blogs, how awkward it can be hanging out with a whole group of people who are nice but are still looking at you like, *"and...what are you exactly?"* Yet they're such good people it's too uncomfortable to say aloud, which makes it even *more* uncomfortable.

Dating someone from a different culture requires us to familiarize ourselves with said culture and, to an extent, adopt some of their cultural traits. Some might not love to hear it, but most of us fit into generalized cultural algorithms. The truth remains that unless a pretty lady of a different race is into ponchos and sombreros if I don't adopt some characteristics from their cultural background, chances are I won't be getting far. It's not about either one changing who we are, unless you really dig theirs, of course, but if you want to expand your cultural dating panorama, you must adopt some other cultural habits/traits; nothing more, nothing less. Something entirely simple.

As you date your potential candidates, remember to select someone who brings more than wealth to the table. Otherwise, you'll be wondering where all the excitement is after a while. If you want to grow together and complement each other, it has to be on an even playing field. Might as well watch paint dry otherwise. Remember our previous chapters and look for the closest man to the overall package.

ELEVEN

The Honeymoon Stage

BUILDING STANDARDS VS HAVING EXPECTATIONS

> ❝
>
> *"Toto, I've a feeling we're not in Kansas anymore."*
> – Dorothy, The Wizard of Oz, 1939

T he restaurants, the laughs, the trips, the gaining weight, the sex—everything is perfect until it's not. This stage in the relationship is full of fun, dopamine, and oxytocin, which is why we feel so excited all the time, and everything seems perfect. After all, we're getting to know someone, and every turn we take seems to be the best thing that's ever happened. Love is in the air, but love is also blind. While this is not the moment to hold back, it's also not the time to form expectations, either.

By all means, indulge away, but never lose sight of who you are and what you want. It often happens that couples get past this stage, and when the rose-colored glasses come off, they often say, *"Well, she wasn't quite what I expected."* A good follow up question would be, *"Well, why did you form expectations in the first place, Sherlock? Why would they be what you formed in your mind?"* As you get further into the relationship, you start to wonder why that person is not the person you envisioned, making it seem as if that person had somehow failed when the failing was done all on your own by building expectations out of your own aspirations. These are more appropriately called *assumptions*.

Building expectations is not wise, because we build them out of pure imagination and use the moments of ecstasy we're experiencing in our honeymoon stage to create an image of the person we're dating into someone who doesn't exist. Having standards is different than forming fictional characters out of thin air and then being disappointed when Wonder Woman was human after all. Having a standard is having a vision, just like we formed the "ideal man" in earlier chapters, and there's absolutely nothing wrong with that whatsoever. If anything, it is healthy to develop visions of any desired outcome. It would be different if you envisioned the ideal man and you were shocked the first guy you went on a date with didn't meet the "ideal man" image.

The same concept applies to the honeymoon stage of our relationships. Take this book, for example. Upon writing it, I have a standard for the quality I want to achieve in the cover, structure, writing, getting my point across to my readers, and the marketing strategy for it to be successful. I have a vision of it becoming a *New York Times* Best Seller; few things would make me as happy. Nonetheless, I do not expect that it will become one only because I hope it will. I employ a very realistic view on the chances it will.

Similarly, in our relationships, building a standard of where we want the relationship to go is something that not only would make us happy if met, but it would even surprise us, whereas an unmet expectation would disappoint us. Overall, I guess you could say it all boils down to a pessimistic vs. optimistic vs. realistic perspective, which makes all the difference. While having a standard makes me work harder and do everything in my power to meet it, an expectation limits my output since it is hope without a factual basis.

We've previously established how irresponsible it is to go into the dating world with a "let's see what happens" mentality. Well, our approach shouldn't be that different when we've jumped into a relationship. As aforementioned, by all means, have your fun, relax, enjoy every aspect of the person you like, but do not be carried and swayed into outcomes. Create your outcomes. Plan some trips, test the waters; after two months or three months, it may be the perfect time to take another step and perhaps meet each other's parents, or plan an international trip, or spend Christmas together.

I'm not necessarily saying to rush into things, no way, but I am saying that if those things frighten you or they're not part of your plan, vision, or near future, well, who are you kidding? Why are you in a relationship? It's quite callous, if you ask me, to create such deep connections with someone to potentially sever them all because "Ms. Daisy" and her casual self didn't feel like "Noah's" emotional state was worth the thought and time in the day. What would you think if it was the other way around? It is irresponsible and immature, and much else I can't say for the sake of the safe publication of this book.

The truth is when going into a relationship, we sign an invisible contract that says you don't get to cop-out and say, "well, that's just not my personality." You enter a relationship with the underlying understanding and unspoken agreement that a relationship is a two-way street, and compromise must be met. It will be work and fun, and there will be ups and downs and much else. We must take the good with the bad and implement some strategies along the way to smoothly sail through this voyage. Otherwise, you shouldn't be in a relationship. In all fairness, it has taken us nine chapters to get to this point, so to simplify the formula—be attentive while enjoying your honeymoon stage and decide whether to pack your bags or go for the long run. Remember, it's a marathon, not a sprint.

Also, as you move forward in your relationship, know that your first benchmark is at the one-year mark. Statistically, most relationships end within the first year, since it's said to be the most challenging year in a relationship.

Also, in case you happen to feel a little more snuggly during July or August, it may help you to know that those are also the times of the year where most relationships start since people are looking for a buddy to hibernate with during the winter. Similarly, most relationships end during the spring. Sorry to burst your bubble, but humans are a bit more predictable than most people think. In that case, you'd be wise to be mindful of the mystic chords of the seasons and the possible temporary effects they may have on us.

A relationship is like anything else you want to get good at—it takes practice. I'm sure any parent can tell you there was a bit of difference between their firstborn and their second one or even their third. The first lucky little rascal always had a hand on his/her chest to see if they were breathing, the second one immediately got baby radios, and the third one was more of a Charles Darwin "only the strong survive" approach. Keep the realistic approach we've maintained throughout the book, as we have three more chapters to get you cruising through your relationship—or make it out in one piece (can't make any promises).

Newton's Third Law of Motion
FOR EVERY ACTION, THERE IS AN EQUAL AND OPPOSITE REACTION.

The energy you put out is the energy you shall receive. Do not morph this into "the energy I shall receive is the energy I shall give." That's regressive thinking, not to mention passive-aggressive. You don't get to choose Prince Charming and think the qualities you seek in him only apply to him and not you. It simply doesn't work that way, and that insensitive manner of thinking tends to compensate you with being alone.

Work is a subject of two, and though work in a relationship can most certainly be challenging, a couple working together can be quite enjoyable as well. It's not necessary to associate work with any negative connotation. Working together can mean fun projects, finances, moving in together, meeting the parents, an international trip together, plans overall.

Once you try it, you'll quickly find they only build upon all elements of a healthy, loving relationship; elements waiting for you ahead to make the best out of your relationship: *The Foundation: Communication*; *The 4 Elements of Love: Respect, Honesty Compromise, & Trust*; and the element of life, *The 5th Element: Sex.*

It is crucial to hold to your individuality and not lose track of time, because God knows time flies when you're having fun. If you hop on this ride without the notion of time, identity, or purpose, well, you'll be hopping off before you can say "blueberry cheesecake" while wondering how you don't have enough to purchase another ticket.

The honeymoon stage is the moment to take notice of who you both are, and also who you are not, and what you like and don't like about each other. Though our vision is towards a successful and healthy relationship, let me slip in this bitter word of caution. In the worst-case scenario, this may also be the safest time to break up if it doesn't look at all fruitful. Otherwise, you're in for the ride, and you should be 100% certain you want to be on it because this ride came with a warning sign before you stepped in that said, "this may be a bumpy ride."

So, what does that mean?

It means, "Buckle your seatbelt Dorothy, because Kansas is going bye-bye."

-   Cypher, The Matrix (1999)

TWELVE

The Foundation: Communication

THE ART OF EXPRESSION AND COMMUNICATION

A s much as we have heard about the importance of communication, we rarely ever hear about the importance of expression. Though both words carry different meanings and have distinct purposes in our love lives, they are one of the same, ever entangled together as two sides of the same coin. They are both essential elements of all successful relationships.

Communication

Communication is the art of information flow. In the specific case of relationships, the art of communication is information traveling on a two-way road, flowing from point A to point B and back. Nevertheless, the art of communication, as we stated in our previous chapters, is as much about *what to say* as it is about what *not* to say, *how* to say it, *when* to say it, and also when *not* to say it. Additionally, silence and space also speak a thousand words, and no, I'm not talking about the silent treatment. I'm specifically referring to verbalized agreements of distance and silence.

Communication is about working in unison by being conscious of the fact that you are two different individuals in every way. You have different upbringings and life experiences overall. It's like tossing different ingredients into a bag, shaking it, and expecting for the marinade to stick to only one side of the meat…good luck.

Communication is about getting to know each other. It is about establishing order among the chaos and also about building lines of respect and setting boundaries. See it as the foundation of a structure. Without foundation, it'd be impossible even to begin.

In a past relationship, I would continuously get backlash when I inquired about why she hadn't responded in hours or asked what she was doing. I would often be labeled jealous or possessive, and I began questioning myself to an extent where I even started to believe it. However, what we fail to remember is that we are, by nature, pattern-recognizing machines, and the only way we can program the machine to recognize the correct patterns is by feeding it the accurate information through communication in the first place. Our brain receives information strictly through our five senses: sight, smell, hearing, taste, and touch; telepathy isn't one of them. So, we can't expect our partner to know what *we* don't openly communicate with them.

Lastly, communicating should always be with higher thought to emotion ratio. Emotion is acceptable as long as we gauge it and control it with our rationale, and not have our thoughts influenced by our feelings.

Expression

On the other side of the coin, while all communication is a form of expression, not all expression necessarily turns into communication.

Furthermore, while communication is merely expressing our thinking process verbally, *expression* can be best described as the manifestation of emotions through various means. Therefore, expression is not bound to only one of the senses, and we can manifest it in the form of kind gestures, moments, words, and more. For example: "I love you," "you look lovely," "I like the way your hair looks," sitting down to enjoy a movie together, or giving or receiving a shoulder rub after a long workday are all different forms of expression. If you've ever read Gary Chapman's *The Five Love Languages,* he does a fantastic job in breaking down what he believes are the most important forms of expression among couples: Words of Affirmation, Acts of Service, Receiving Gifts, Quality Time, and Physical Touch.

Expression is a one-way street, and it's essential to understand that expression should be selfless and be accomplished without prebuilt expectation, even though we usually misinterpret its purpose. You should express your emotional language purely because that's how you feel, not because that's what you want to hear back.

We often make the mistake of expressing ourselves with prebuilt anticipation of a reciprocal response, which is our mistake, not our partners' in not responding with a similar remark. Anticipating a response can lead to the buildup of questions born out of insecurity and, with near certainty, ultimately create conflict. So, does that mean we shouldn't ever expect a response? Not necessarily; it means you shouldn't expect it when you're expressing your feelings. Yet, if you don't ever receive expressions of love from your partner, that's something that you should ponder and *communicate* with your partner.

Live by a philosophy of accepting only positive energy into your life, while working on the negative or discarding it entirely from your life. Practice the real power of "live and let love."

Communication Acumen

Annie: "Where did you guys go?"
Becca and Kevin: "Disney World."
Annie: "Oh!"
Becca: "We finish each other's sentences. Sorry!"

– Bridesmaids (2011)

Have you ever had a moment like that? Do you know a special someone with whom you share a connection where you finish each other's sentences? That's what you call a special bond.

We've all been to social gatherings, gone on dates, and met new people. All of those activities have something in common—they require at least a sense of social acumen. You might be a great conversationalist yourself. However, communication is more than a good conversation, it is more than small talk, it is more than socializing.

Communication is about connection and intimacy; it's about getting to know the person you're with on a deeper level, it's about finishing each other's sentences. Communication is about the little things as much as it is about the bigger things. It is about the laughs, the inside jokes, it's about understanding what makes the other person tick and what makes them smile. It's knowing their ticklish spots and when they need to be left alone. Over time it's indeed about believing telepathy is a sixth sense, about getting to know them as much as you know yourself, not that you are a graduated self-connoisseur. But more importantly and opposed to common consensus, it's not so much about talking as it is about *listening!*

Have Discussions, Not Fights

Are you familiar with the quote, "Don't raise your voice, improve your argument"? You are now, and it holds quite a bit of weight. Like most quotes, it operates by highlighting the irony in the actions we take through life's journey.

Yet, in the fun and crazy world of relationships, disagreements are, and will forever be, as certain as death and taxes. I'm sure you've heard of the famous idiom, *Life isn't about waiting for the storm to pass. It's about learning to dance in the rain.* This basically means we need to learn how to cope with the things in life we can't control, such as weather, time, and this case, disagreements in relationships. So in the same perspective, it's not about having or not having conflicts; it's about how we deal with them.

As with other subjects, your emotions are no more than a cocktail of electrochemical signals communicating throughout your body. As we've learned, these signals are not always positive. Sometimes we feel sadness, other times we feel frustration, and some other times even anger. Experiencing those emotions is what makes us human; controlling those emotions is what makes us mature adults. And sure, you're human, after all; I'm not exactly asking you for perfection here. Regardless, one of the main components in having effective and healthy communication is learning to have discussions, not fights.

Whether the disagreement is about something significant or something small, we often fail at finding or losing, for that matter, the purpose of our interaction. This leads to the infamous arm's length competition, causing us to lose sight of the issue at hand, somehow managing to morph the original point of having our conversation in the first place to a battle of who's right, ultimately creating what we otherwise know as conflict.

Two Negatives Do Not = A Positive

If you're a math geek, you might be whispering to yourself, "Hmm… You're mistaken, actually; two negatives can equal a positive." Hands off the trigger, this is a class about love. In all seriousness, we have *all* been guilty of this at one point or another. Your partner reminds you of washing the dishes or groceries, etc. since it was your turn, and you counter by reminding them they forgot last time too. What we fail to see is the underlying question here: *what does it solve?* You guessed it—absolutely nothing. If anything, it only adds fuel to the fire—and wood—and fireworks—and uses kerosene as the source of fuel while it's at it. So, no bueno!

If this book is to underline anything at all, it could best be summarized by our overwhelming human capability at operating in a constant state of emotion. Most issues we experience in relationships are self-made. The reality is that real problems are not as expected, and when you encounter one (God forbid), you'll be quite aware of the severity of the problem. Otherwise, couples fail to get along at the astronomical rates we see nowadays because *we* don't allow it to work. Our unwillingness to put in the effort and express our vulnerability is nothing short of baffling. Like saying we want to open the door, but not getting up to open it, and then having the nerve to express complete astonishment when the door remained closed.

Tip # 1: Gauging Defensiveness

Answering a wrong by pointing out another one is being defensive. People usually revert to defensiveness to protect their insecurities from information that threatens them. If you're confronted with a similar situation, take a deep breath, put into practice some of the key takeaways from Dr. Christian Conte—swallow your pride, and activate your thinking brain to assess the info given by your partner more maturely. You'll soon find out the severe consequences of admitting you're wrong are absolutely none. Eventually, the more you put this into practice, the more you'll realize how much time was wasted with the contrary.

Tip # 2: De-escalate the Situation

Every object in a state of uniform motion will remain in that state of motion unless an external force acts on it.
– Newton's First Law of Motion

If you're the one waking up to a dirty kitchen when it wasn't your responsibility to clean it and upon confronting your partner about it, you hear a loud noise coming from his mouth resembling a bark, nip it in the bud. Be the bigger person and show him the power of control. Instead of counterattacking with an outburst of your own, reaffirm and redirect the flow of the confrontation by saying something like: "I acknowledge I didn't wash them either and will do my part not to repeat it. I'm not saying you're irresponsible, I just want us to work together on our commitments, that's all."

If your partner still has his guard up after a similar statement, do yourself a favor, take a page from Newton's first law and remove yourself from the situation. Your partner is currently in an emotional state, and if you remember correctly, emotion is energy in motion. So the worst thing you can do is add more energy to theirs. Even if you are annoyed yourself, if you *genuinely* want to take their power away from them, let them deal with their own storm. There's no benefit in jumping into the eye of the hurricane yourself. Negative energy is highly draining, and who we choose to give it to and how we choose to express it is ultimately up to us.

Tip # 3: The Fortune 500 Strategy

Some of the world's top companies implement similar strategies based on psychology to handle their angry customers. Take, for example, this company's strategy to defuse an angry customer (whose name I can't disclose for obvious reasons):

Listen → Acknowledge → Empathize → Present a Solution

1. **Listening** – Let the person vent as much as they please. The philosophy is the same as the one above. Adding energy to the situation will only worsen it. Instead, let them release it completely. No one with a healthy mind can really remain angry on their own.

2. **Acknowledge** – Use your active listening skills, make sure you hear the true issue, and acknowledge the reason your partner is upset. Paraphrase it back to them to show them you've listened and gain their validation by saying an agreement statement that puts you both on the same page: "Did I get that right, hon?"

3. **Empathize** – It is essential to feel your partner's pain or frustration during a difficult situation, to an extent. Always remind yourself how you would feel if you were in their shoes.

4. **Present a Solution** – Lastly, being able to provide a new direction or solution is valuable. Shifting the focus to what you *can do* is more productive than continuing to focus on what happened in the past.

Two Rights Can Co-Exist

If I told you 2+2 = 4 and you were to correct me by saying that 2x2 = 4, those would both be accurate statements. We often ignore the fact there are different routes to arrive at any given solution. The fact that I stated a fact about the same subject as you doesn't have to negate the validity of your point. Indeed, two right answers can co-exist in harmony. Yet, like in most cases, what happens is that our insecurities take the wheel, and to avoid crashing, we hyper-focus on not being wrong, instead of having direction.

As a result, the validity of being right becomes more important than the truth of our message.

Some time ago, some friends and I were hanging out at a local bar, enjoying some good drinks and grub, and having some casual conversation. Before the night was over, two of our friends started arguing about the struggles of life. One of them stated certain people faced harsher obstacles in life than the rest. Friend #2 felt offended and claimed he had paved his way and that the luxuries in his life were a product of his hard work, to which I interjected in their conversation by saying: "Be that as it may, a point can often be argued both ways. And just like it's true that you paved your way, and no one denies your struggles, in the same sentence, it is also true that others have faced harsher circumstances."

And ta-da, problem solved.

Apologizing Like a Pro
"I offer you an apology..."

Oh, gee, thank you, I've needed an apology for a while to put in my wallet. You never know when it might come in handy. Have you heard an apology that for some reason, you couldn't explain why, you were still upset after hearing it? Perhaps you felt confused as to why you were still upset, followed by a sense of guilt because you were, and also not understanding why. That's because you were right to feel the way you did. You didn't fail at accepting an apology; the person offering it didn't give you a proper one, and your subconscious brain picked it up—leaving you with a feeling of yuck.

To understand what a proper apology consists of and the real purpose behind it, we must understand what words are first. We can trace the origin of written words back about six thousand years. On the other hand, the manipulation of soundwaves, otherwise known as language, has been around

50,000 to 150,000 years. Quite the gap, huh? That's pretty much the entire span of modern human existence. Long story short, we don't know with certainty, and the only evidence we have even to suggest that timeframe is bones found in congregations, meaning we learned to speak when we met the need to communicate in groups.

Fascinating stuff, right? What that suggests about our use of words is that we invented them as a means to describe our actions. What happened is that as thousands of years passed, turning memory and origin to dust, we did exactly that—we forgot. Yet, your brain remembers. The nature of communication is so engraved into your subconscious mind that a baby will attempt this skill by babbling as young as four to six months. To this day, that is the purpose of language; words are representative of actions, not a substitution, hence why you remained upset.

In the most literal of ways, when you offer an apology, the word "apology" on its own has no substantial value. The person hearing the apology has no actual use for it unless it's tied to a described action. Might as well tell someone to extend their hand and place a rock on it, and don't be surprised if they throw it at you—which is actually more than you can do with an apology, hence the frustration. In essence, it's similar to when you're upset or sad, and you can always count on some genius saying: "Hey, don't be sad, you should feel better." *Oh, hey, thanks, I never thought of that option, what a relief.*

As well-intentioned as their statement may be, the truth is the words in the phrase itself have no value as they are not tied to any action in the form of an instruction or actual meaning. The point is that words are being used as substitutes of action, not representing it. Precisely because of that, the body of every apology should be broken down into three parts: *recognition,* an *offer,* and a *promise.*

It is also absolutely vital to note an apology should never under any circumstance include an excuse, which is usually misinterpreted by the person delivering the excuse as an external "reason" they failed at their commitment. Excuses are like a*****es, everyone's got one, and they all stink. Also, as I stated before, actual problems are few and noticeable when they indeed show up in our lives, as they are external circumstances we *can't* control *nor* prevent. If the event that impeded you from your responsibility was something *entirely* out of your control, there's no reason why your partner won't understand.

You are doing yourself a service by eradicating excuses from your life as they are but a figment of our imagination; they are not real.

Recognition – This part of the apology is about recognizing the problem and tying it to the action that's making you apologize in the first place. For example, if you and I were dating—don't laugh, I'm serious—and we had planned on an evening together where I would pick you up at 7:00, and I didn't arrive until 7:34, it's safe to assume you wouldn't be the happiest camper. In that case, the recognition part of the body of the apology should sound something like this: "I'm sorry I'm so late to pick you up. I have no excuses, I'm late, and I understand why you're upset."

The Offer - The offer is the verbalizing of the apology, which is more than saying, "I'm sorry." If you told me about your bad day at work, and I told you how sorry I am you had a bad day at work, I'm not necessarily apologizing for your day, am I? So an apology is personal. As the name of this chapter suggests, it is about communication, about tying what I'm sorry for to *me*. Like the other two parts forming the overall body of the apology, offering an apology is an essential part of showing ownership of that for which we are sorry. It would look something like this: "I offer you a deep and sincere apology for being late."

The Promise – This part of an apology can be explained as a summary of an overall solution. The promise relieves the doubt of the thing you're apologizing for recurring in the future. In other words, the overall body of the apology not only addresses the past action but also alleviates the concern of it repeating while showing an effort to change. Which, in its entirety, communicates more than the context of the words expressed in the apology. As opposed to saying "sorry," which implies less than the context of the words said. The promise sounds like so: "I promise to work more on myself moving forward, and it won't happen again."

As a brief word of advice, if you're apologizing, try to refrain from saying "try" in your sentence. It programs our brain for failure, sending the message into our subconscious that if the word exists in our attempt in the first place, then the possibility of failure is most definitely present, consequently making failure evident and unavoidable. For the sake of the example, if you were in a dance competition, you wouldn't say—or at least I hope you wouldn't— "I'm going to try to win" or "I'm going to try to be good." Otherwise, why compete in the first place? Yet, if you're not comfortable overall with selling complete certainty in your approach, then replace "try" with "I'll do my best." Not only does it have a better ring to it, but it embodies the effort and commitment that both of you as a couple deserve.

The overall apology looks like this: "I'm sorry I'm so late to pick you up. I have no excuses, and I can understand why you're upset. I offer you a deep and sincere apology for being late, and I promise to work more on myself moving forward, and it won't happen again."

The Communication Blueprint

When we take our phrases from above into perspective: *what* to say, what *not* to say, *how* to say it, *when* to say it, and also when *not* to say it, what do they mean? It's important to understand the overall description is in no official order whatsoever. We'll assign them by steps in an orderly fashion. The general blueprint can be best broken down into two sections:

Approaching the Conversation

Considering influencing factors before engaging in a conversation is as important as the conversation itself.

Time – *When to do it.* Preferably, choose a time where you both are free and at a time where stress is most likely low is ideal. The morning before heading off to work is hardly the appropriate time. Perhaps a weekend or a relaxing evening over a glass of wine.

Emotions – It's important to be honest with ourselves to understand our feelings better. Having emotions is okay and normal as long as we have gained control over those emotions. An emotion guiding action is called an impulse, and it's the root of all arguments. An action by guided thoughts while having feelings is called a discussion, and that's precisely what you want.

Understanding Worldview – Daniel Patrick Moynihan once said, "You're entitled to your own opinions. You're not entitled to your own facts." A worldview is the fundamental cognitive orientation of an individual encompassing the whole of the individual's knowledge and point of view. In other words, our life experiences shape our perspective of things and also our understanding of things. So when approaching a conversation, keep in mind your partner's brain will decipher information differently than you. Simply try to be comprehensive about that.

The best way to approach their subjective perspective is to acknowledge it, and two, try to communicate in a way that will best fit their personality. Therefore, try to imitate them—some people prefer detailed explanations, while others simply want to hear the facts, and others are more in touch with their emotions or are analytical. Determine their language, and you'll see it's a breeze.

Don't Form Assumptions – Refrain from forming scenarios in your mind. This is literally the purpose of talking in the first place. Remember Mark Twain's quote, *"It ain't what you don't know that gets you into trouble, it's what you know for certain and it ain't quite so."* Misinformation is way more dangerous than a lack of information. This goes hand in hand with not making accusing statements. Have a mentality of solving and discovering by admitting the naked reality—you *don't know* what's in their mind or what they feel.

Accusations lead to fights and naturally cause the person to become defensive. And rightfully so. Have you ever heard someone say, "Don't back a dog into a corner"? Cornering your partner leaves them without options for actual engagement. The only road is to object to your statement, and by doing so, you'll successfully be engaging in a most self-fulfilling paradoxical catastrophe.

Unpacking the Conversation

When I talk to friends and acquaintances about love, I notice how much anxiety can surround addressing an issue. There's no need to complicate it. After going through the two-step process model, take a deep breath, smile, and manage all the problems by saying, "Hey, I would like to talk to you."

As a small piece of advice, using "you" in your sentence rather than "us" or "we" is a way to let them know the focus of the discussion and responsibility of the issue pertains to them specifically.

I. **The Message:** Output

Think about the actual message you want to convey. This part of the message is the information and data you want to relay to your partner. Every message has four facets, according to Friedemann Schulz von Thun (1981) and his Four-Sides model of communication.

1. *Identifying the fact* – A fact is a thing known to be consistent with objective reality and can be proven to be right with evidence. This is exactly that part of the message that has the data you want your partner to hear. In other words, what are you informing him about?

2. *Self-discovery* – What does the message say about you? What does your message make you think or feel? Suppressed feelings aren't healthy. They're good at creating conflict and can often lead to passive aggression.

    Instead, be honest about what you feel and express it firmly. Assessing these inner conflicts are all part of the process of getting your partner to understand you better. Furthermore, describing how you *would like* to feel is as important as expressing *how* you feel in that present moment. For example, if your partner mentions his ex a lot, you might want to say: "I notice you mention your ex a lot, and it makes me think she might still be important in your life, and that makes me feel jealous and unappreciated. *I'd like to feel important in your life.*"

3. *The relationship* – Communicating how the underlying issue impacts the relationship overall is vital to add emphasis and highlight the

importance of solving the subject. Take as an example the message above and factor in this point: "I think it's affecting our relationship, and I'm afraid if it continues, my emotions will only heighten and eventually grow resentful towards you, and I would like to avoid that."

4. *Influence* – This is the sales pitch, the close, the essential part of the message that persuades the listener to change. It can be something as simple as persuading them by saying their effort or change would make you happy. Using something small such as not washing dishes as an example, you could say, "By doing so, we'd have more time to focus on quality time." Or, "it would encourage me to do more things for you." If you choose the latter, remember to thank me.

II.   **Presenting A Solution**

In business, we often say, "bring me solutions, not problems." The phrase is stating the obvious; it's emphasizing the fact that solutions are ideas that present benefit to where a problem by itself is, for lack of a better term, whining. It adds to the pile of a thousand other issues and complaints and further complicates the overall complexity of this thing we call life.

It's also fair to approach the conversation admitting you don't know how to solve the issue on your own and would like your partner's input to tackle the problem together. But washing your hands like Pontius Pilate, alleviating yourself from all responsibility, and shifting the whole weight of the problem onto them might be quite a convenient mentality for you to adopt. Yet, it is not only unfair, but it is also a potent concoction for disaster.

Therefore, think critically and ask yourself how you would want the underlying issue to be resolved. Doing so will present options to your partner, making it easier for them to accept your overall point while making it easier for you to obtain what you want out of the situation.

III.   Listening to Their Response: Input

"... and the last time this happened..."

"But I already apologized for that time. You know why I was late. But go on."

"Hmm, I forgot what I was saying..."

One of the underlying contributors to heated arguments is built-up frustrations due to constant interruptions between the two people having the discussion. Be respectful of their turn, don't interrupt, and engage in active listening and analyzing their response. Even if you've gotten better, you don't get any brownie points for interruptions, but let's not chastise you, either. In all fairness, you're not alone if listening is not exactly your forte, but it shouldn't be an excuse either. The truth is that most people are not good listeners. Most of us would rather be heard than listen to someone else talk.

I'm sure you've heard, "practice makes perfect." Well... I don't know that perfect is the exact word I would use in this scenario, but I do have some good news for you and can promise you'll become much better if you practice. So, try the following with a good friend or family member. It doesn't necessarily have to be your significant other. Still, if they're willing to, that's even better, as it speaks a ton about your relationship, besides making the exercise more fun and productive.

- The rules are simple. First of all, put all distractions aside. That includes your phone and having it facedown doesn't count. Next, grab a notebook and pen and set them by your side, as you'll need them later. Also, try to keep your hands visible at all times. No, this is not a traffic stop, and it's not to be sure you won't stab the poor bastard. It's because fidgeting, grabbing things, or playing with our hands takes our focus from what's being said by the speaker.

- Next, genuinely ask them about their day. Let them speak for ten minutes without interrupting them whatsoever. An excellent tip to keep in mind that will help you focus on the content of what they're saying is making good eye contact. Besides, they'll feel appreciated by your engagement.

- That wasn't that hard, right? Fantastic, but not so fast, slick! After they finish telling you about their day, grab the notebook and pen you have by your side and write down with as much detail as possible what you recall about their day. Try this one to two times a week, and in no time, you'll see you become a great listener. The exercise is meant to teach you patience and to build lines of respect, which will benefit your relationship in an array of ways, from your partner feeling comprehended, to amplifying your own level of understanding and improving your content-retention rate.

IV.   **Resolution**: Making A Decision

Reaching a resolution is simple. You've already put in the hard work. On the other hand, don't get me wrong, making a decision is not easy, especially when talking about love and relationships, which encompass your personal life in so many different aspects. It's recommended you base your decision by breaking it down to four options.

1. They are receptive to your message and accept the solution you propose.

2. They are receptive to your message but propose a different solution, and you accept it since you are satisfied with the outcome.

3. You compromise because you have different viewpoints and, unfortunately, aren't in agreement. Understandably, this outcome may bum you out since it's not the desired result you were hoping. But as long as you arrive at this solution by answering the question sincerely: *"Am I wholeheartedly willing to be part of this compromise, and will I continue to be happy in my relationship having done so?"* There's no fire, so take your time answering. If your honest answer is yes, great, I'm happy for you. Now it's important to take ownership of your decision and indeed let bygones be bygones and move on from the subject to continue enjoying your relationship.

Precipitation can lead to your unhappiness and settling for less than you're worth by being unkind to yourself, and in many cases, rushing into answers by sheer impulse is the road to the valley of unnecessarily ended relationships. It's okay to say you want to think about it. You have the right to be kind to yourself while being kind to your beloved. It's as simple as saying, "May I take some time to think about it, please?"

4. End of the road. If, upon asking the same question from above, you answered no, then ending the relationship might not be the desired outcome but, unfortunately, the healthiest route for you.

Remember the work you put into yourself to get here in the first place. In the previous chapters, we learned the importance of loving ourselves first before jumping into a new relationship. So staying true to that promise and the person in the mirror couldn't be more significant. Recognize your partner's right to their own opinion and point of view, yet don't allow the fear of losing someone special to you decide for you.

With full optimism, I hope you both reached a happy conclusion, and you get to join me in the next chapter.

THIRTEEN

The Four Elements

RESPECT, HONESTY, COMPROMISE, AND TRUST

T he ancient Greeks believed everything on the planet was composed of four elements—earth, water, air, and fire. Later, Aristotle added a fifth element he called the aether, which was the invisible element that filled the rest of the empty space of the universe. For thousands of years, these elements were the foundation for the pillars of science, philosophy, and medicine. Similarly, here are the first *Four Elements of Love: Respect, Honesty, Compromise, and Trust.* Plus, a fifth and essential element to love and life: *Sex.*

A relationship is bringing together two perfect strangers with different lifestyles, who often have different cultures as well. Both of their experiences have been shaped by countless memories that make up their personalities to perhaps one day form a family. Therefore, it is by no means easy, and it will take commitment and an array of different factors to make it work and enjoy it to the fullest. Hence, why these elements are *not* the only components your relationship will need, but instead, the elements no relationship can live without.

They are also in order of structure as a building block that supports the other, not importance level, nor a block that can exist without the other. Having said that, after our foundation, communication, it takes **respect** to form **honesty**; it takes respect and honesty to be able to **compromise**; and it takes respect, honesty, and the ability to compromise to build **trust**.

Respect

Self, Partner, and Relationship

If I don't respect your integrity, it's going to be awfully hard to trust you. If I can't respect your judgment, it may be hard to compromise with you. If I can't respect your self-awareness, it may prove challenging to be honest with you. And if you don't respect yourself, how can I respect you?

Thus, as you have been practicing for the majority of this book, the first stage of respect is *self-respect*. Yet, if we can't respect, hold accountable, be fair, love, and stand our ground with the person in the mirror, why should anyone maintain a higher standard for that image than the one we hold ourselves? They shouldn't, nor will they. I'm afraid to tell you that if you don't have the strength to respect yourself, no one is going to do it for you, and if you are willing to let someone disrespect you, it's going to be rough.

For example, let's say someone cheats on you, or you find out they were in a relationship with someone else while they were with you, and for whatever reason, after talking about it, you're willing to forgive them and give them a chance at "redemption." Though change is certainly possible, it becomes nearly impossible when we've communicated to that person not only that we're willing to be disrespected, but how far we're inclined to be disrespected.

You may want to look into "The Stanford Prison Experiment," which remains one of the most controversial psychological experiments ever performed. The study claimed to prove the root of evil and how with positions of authority, such as in Nazi Germany, prison systems, and more, most people are very capable of inducing harm upon others once they've been allowed to see the measure of pain they can inflict.

This theory can be applied to a reasonable extent to many other aspects of life, including our scenario here when talking about relationships and infidelity. As human beings, we have invisible stop signs for our violence, our corruption, our kindness, and nearly all else.

Take someone who claims to be overly aggressive and acts up with his family, girlfriend, and even friends. Yet, one day in front of police officers or when faced with an overpowering opponent such as a towering, six-foot-five, strong male, his aggressiveness seems to find control. That's because, in most cases, it always has a level of control. It is our own allowed subconscious perception of what we're willing to do or not in a given situation. In other words, the person is unlikely to have a real aggression problem; they have just excused themselves to a stop sign that allows them to inflict aggression upon those with whom they feel comfortable.

In other words, the unfaithful partner is unlikely to feel genuine remorse. They've just met a situation that triggers the understanding they must act accordingly, because you still have something they don't want to lose—whether it be assured sex, nurturing, financial comfort, or something else. What's worse is that the stop sign of how far they're willing to go disrespecting the relationship has just been moved and allowed a broader parameter by you forgiving them.

This example is why personal standards should be met, and boundaries should be set, because once someone has crossed your lines of respect, make sure to remember they'll cross them again. It's quite simple—you have helped them redefine their perception of your worth, as where else could forgiving unfaithfulness come from if not from fear and insecurity (excluding marriage or having kids together)? Forgiving infidelity often comes from fear of being alone and insecurity of not knowing our capabilities for attracting a good match out there. It may sound a bit drastic, but disrespect should be enforced with a zero-tolerance mindset.

By forgiving their sexual "mishaps," you have officially and successfully let them into two grave little secrets—one, you're an option to them, and two, you don't have any. The grotesque tone of that statement hails in comparison to what awaits when you allow your partner to step over these lines of respect. *I love him—we've worked it out—he knows better—I got him back*—sound familiar? They are all excuses from the wounded soul of the gorgeous person inside, asking you to help her.

Human beings are capable of beautiful things, as they carry a lot of beauty within. The human mind is capable of creating beautiful things; people are capable of feeling deeply for one another and of making a lot of good. But we've often experienced so much of each other in high dosages that it has numbed us to the core and built walls to keep us safe from each other.

Whether it's our biological nature, our surroundings, or whatever the case may be, it's irrelevant at this point on our venture. Still, what is true is that kind people often have a higher respect for others than for themselves (high on agreeableness). Though the world needs more people like them, they often get taken advantage of by the very people they're kind to and overly respectful. Therefore, when you've entered into your new relationship, remember to respect yourself first and then people, and your partner will respect you.

One of the most important aspects to keep in mind about self-respect is that respect should be neutral—not defensive, offensive, aggressive, and least of all, regressive. If your self-love and self-respect spark any of those mentioned actions, then it's as simple as taking a closer look at your current definition of self-respect and making some tweaks. It's hard drawing out a purposeful example. Nonetheless, respect is about having standards and setting boundaries from defining who you are and what you want, while remembering this is not the same as building expectations.

Similarly, if you can't respect yourself, how can you ever expect to *know* how to respect someone else? Therefore, the second and equally important type of respect in a relationship is respect towards your partner. This factor is, unfortunately, where the problems often lie and equally as often could have been resolved if we had just asked ourselves two things: *Would I expect from myself what I demand from my partner?* If the answer is yes, okay, now communicate it. Secondly, *would I like or be okay with what I express or impose upon my partner if it applied to me?*

It is also essential to understand that even those two questions won't always resolve issues in our relationship since our partner is someone with a complete, individual view on life, values, and wants. Therefore, respecting our partner also means being respectful of their thoughts, likes, and dislikes, as they may be different from our own. It's acceptable to go back to our communication chapter to freshen up on some key points.

For example, an ex-girlfriend of mine would often make unsolicited comments about her past sexual experiences. Her comments left me confused and upset as to why she had mentioned it in the first place. The *purpose* remains a mystery to this day, but I suspected the *reason* was one of three possibilities, if not all three. Insecurities of her worth made her validate her importance through sexual encounters. Another possibility was that she suffered from passive-aggression from past trauma that she displayed as a non-direct expression of disgust towards sex. Or perhaps—she thought she wouldn't be bothered if I expressed something similar (unlikely).

It's crucial to be respectful and mindful of each other's differences and past experiences to let the past stay in our past. There's no need for our partner to know about our many endeavors. Trust me, they're adults. They know. Just because someone knows something doesn't mean they want to hear it. Remember to be aware of their personal views and also know that for us to receive the respect of others, we must be willing to give it first. So giddyap on learning about each other.

It's also important to be mindful of our relationship and hold it to a standard of respect. Well, am I not being cognizant of my relationship by being respectful towards myself and my partner already? In many ways, yes, and in others, not quite so much. Sometimes it's easy to be faced by smaller situations where we weren't quite sure what bothered us since our partner was respectful of us as much as they were of themselves. So, what troubled us then?

Let's just say for the sake of the example, that you're at home, and you happen to have a little more work than usual. Therefore, you need to text your boss or team a bit more on this given afternoon. Now before we continue with our point, let's put the following into perspective—a human mind is a pattern-recognizing machine, and muscle memory is a real thing. Otherwise, you wouldn't be able to talk while moving your hands, wake up at certain times without an alarm, nor would we refrain from signaling when changing lanes, since we wouldn't have learned that turning on our turn signal light meant the driver in that lane was going to accelerate.

In relationships, these patterns are even more evident despite us being aware of them. Consequently, we naturally learn our partner's routine and habits despite us not being psychotically neurotic about it. Now, given our new perspective and going back to our example where we have to text our work team a bit more than usual, our partner asks us who we are texting. Their inquiry is not necessarily because they are jealous, untrusting, or suspicious of us; it's because they are human and recognized a deviation from the pattern they are used to seeing. More importantly, it's not that they're disrespectful, but that it takes respect to build trust in the first place. I happen to have been through a similar experience, and my inquiry was confronted with an aggressive "Why?" *Geez, who did you date?* Her response made me feel wrong for asking in the first place since I thought her aggression was unwarranted.

So, in small moments like in our example above, it's as easy as saying, "Hey, hon, we have a bit more work than usual, so I have to text my boss/team to give them an update." I'm sure you'll be met with an "Okay, hon, thanks for letting me know." Being mindful of our relationship and knowing what small acts may build those stronger bonds of affection.

Like in most of the lessons in our lecture, there's plenty of examples of respect between a couple, such as not forming habits out of using cuss words towards each other even if you're joking (occasionally, it *could* be fine). This habit can create vicious patterns axing away at love regardless of both partners being "okay" with it. Now whether we're talking about self-respect, respect towards our partner, or respect towards the relationship, no one is telling you to bark at your partner as justification for loving yourself.

No one is saying you can't express parts of your past to your partner either, or that you have to make them aware of every time you send a text. It's as simple as being mindful of unexpressed feelings or thoughts or the implications we may make upon our relationship by remembering to ask ourselves, *"how would I like to be responded to* and *how would I like to be treated in this given situation?"* Whatever answer you arrived at, as long as it was genuine, do that.

Honesty

When looking back at our foundation chapter, communication, we learned the importance of expression in a relationship. Honesty is one of the critical elements in that foundation of communication. As our partner is getting to know us better, it's easy to get involved in a cat-chase-mouse guessing game because we find it "romantic" or empowering. While this can be harmless in small doses, such as special occasions, allowing ourselves to be captivated by our illusions of grandeur could be surprisingly detrimental to our relationship in the long run.

A good friend of mine once narrated to me an interesting anecdote. She used to work at a manufacturing company where she was the head of the entire supply chain management. She told me how they had a vast number of procedures and used quality-control machines and systems to facilitate those procedures. Yet, what sometimes happened is the systems or appliances would miss an issue with a product. This issue occurred because there had been an error when programming the machine/system in the first place.

Consequently, the machine had no way of knowing it wasn't checking for the issue correctly. She then applied the example to people, and how people often make mistakes or decisions based on erroneous information, hence remaining clueless, that wasn't the correct way to go about things in the first place.

This analogy can also be said about honesty in our relationships. If we're regularly playing the "I'm interesting, *guess* to know me" game instead of a *"get* to know me" approach, or on the other hand, simply giving incorrect or partial information, how can we possibly fathom why our relationship is failing later down the road?

At first glance, being honest in our relationship sounds like a no-brainer, but the truth is that practicing honesty or accepting someone is honest with us takes practice. Most people don't like hearing utter honesty because it's not always positive, and on the other side of things, most people don't practice being honest because they don't like confrontation, and who does?

But even still, avoiding confrontation makes sense to an extent due to people becoming more and more easily offended with time, making it harder for people to practice honesty and overall creating a vicious cycle that is hard to break.

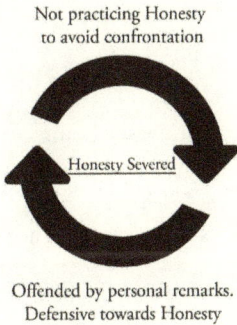

Not practicing Honesty
to avoid confrontation

Honesty Severed

Offended by personal remarks.
Defensive towards Honesty

Yet, getting in between these arrows and allowing honesty in our relationship by practicing our communication skills from a previous chapter, which is both listening and expression, does three essential things for us, while avoiding a dangerous one.

Good news or the bad news first? We'll get the bad news out of the way first. Allowing this vicious cycle to continue can lead to something unhealthy in every way possible—passive aggression. Berit Brogaard, D.M.Sci., Ph.D. writes for Psychologytoday.com the "5 Signs That You're Dealing With a Passive-Aggressive Person":

1.   The silent treatment

The passive aggressor will punish you with silence to avoid confrontation and have you play a guessing game. At the same time, they are prolonging the silence when you don't play the game or guess incorrectly. *Oops, thanks for playing, try again!*

2.   Subtle insults

Your partner has been complaining about you leaving your clothes scattered across the bedroom, and one day you decide to surprise him to picked up the bedroom to put your improvement on display. But when your partner walks into a clean room, instead of giving you credit for it, he gives you *diminishing* credit. *"Oh nice, you **finally** did a good job, hon."*

3.   Sullen behavior

It can be a bit taxing (to say the least) being around someone who is gloomy, thoughtful but inexpressive, or defensively quiet all the time. Make no mistake about it, most likely this is a conscious action, just like when you are gloomy, and you understand it's not your partner's fault is a deliberate action as well. A passive-aggressive person will purposely let you feel their mood and let it damper the atmosphere. Nearly as a provocation for you to comment or inquire about it. Let's say you're hanging out with some colleagues or friends, and "Aron" makes a harmless joke about the intense heat outside. As it's to be expected in social gatherings, everyone laughs, but Janine stays noticeably quiet. Janine is purposely stating her discontent with Karen to make her uncomfortable.

4.   Stubbornness

Being stubborn and being persistent or standing your ground are two very different things though they may appear the same in conversation.

Being stubborn is being unyielding, whereas being persistent is being determined. While even a level of stubbornness can be positive in different life situations, the passive aggressor will do it for the sole purpose of annoying you or those around who are listening. Playing a *causing irritation* vs. *being correct* game—and *what's worse* is that they'll defend their point intelligently, making it hard for you to prove yours.

Also, stubbornness rooted from passive aggression will play a fight not to lose, instead of discussing to prove a point. Continuously dancing around the issue or question posed. Like a chess player who keeps moving their King as the sole remaining piece, instead of giving up. Thankfully in tournaments, that's declared a loss, despite the other player being unable to checkmate (if they have more pieces on the board).

5.   **Failure to finish required tasks**—procrastination

This is not your typical procrastinator who simply decides not to do a task based on many reasons—from being lazy to avoiding the responsibility of doing a job because of the work required to perform it in the first place. On the other hand, the type of procrastination that a passive-aggressive person will show is a stubbornness to follow instructions, despite how correct or beneficial the instruction might be.

For example, your partner is traveling over the weekend to visit their parents, and you've asked them to let you know they made it home safe, but they fail to do so. Well, what's wrong with that? On most occasions, the answer would be absolutely nothing—people forget, as simple as that. But the passive aggressor is different. For one, their "reason" for failing to do so will be unapologetic, defensive, or point elsewhere as the responsible factor. "*Well, it was busy at the airport and I had a lot to do, what do you want me to do?"* when a simple, *"Sorry, it was busy at the airport, babe."* would have sufficed.

This behavior is a death sentence to a relationship. It is toxic, as well as hard to recognize or confront. Since the person inflicting the passive-aggressive behavior will **always** avoid confrontation or ever admit the issue truthfully. If they say a sly remark, upon being confronted, they'll play victim, shifting the problem to you *"oh gosh, why are you getting so upset, I was simply asking a question."* This person involves passive-aggressive behavior to "avoid" conflict/confrontation in the first place and is easily offended, ironically creating extreme conflict themselves firsthand.

On the more optimistic side of things, practicing honesty in our relationship does three very important things: one, it saves time by getting straight to the matter at hand. Two, honesty solves problems since it reveals the true nature and the root of what we need to solve. Three, honesty creates stronger bonds in the relationship that allow us to build trust and compromise, which are two of the four fundamental elements in this chapter towards creating long-lasting and healthy relationships.

Compromise

Not Sacrifice

I think it's fair to say that at some point or another, we have all heard that you must make sacrifices in marriage, and I won't argue against that. But relationships are a bit different. Don't get me wrong, by all means, a relationship should be taken with the responsibility of being with that person permanently; otherwise, why be in a relationship? Still, the duties of an ideal relationship before marriage are not even in the same realm of comparison. In marriage, you may have children or fur children, a mortgage together, bank accounts, etc. Decisions such as outings, sex, or vital life changes such as job accommodations need to be made as a team.

All of those examples are understandable and, to the extent implied here, necessary. Nevertheless, the story is a bit different when we are in a dating relationship, and we don't have that level of responsibility and commitment towards each other. In a girlfriend-boyfriend relationship, making sacrifices can have the opposite effect and impact the relationship negatively. With men, it can turn into becoming a "yes man" situation before you care to notice, which will most likely lead to his partner losing respect for him. On the other hand, if a woman becomes a "yes lady," that could lead to her partner taking advantage of the relationship and her personality. A dating relationship will still have plenty of disagreements, and after communication and even after using all the tools we have covered in this book, we may very well not always resolve the problem at hand. In these moments, we must learn to compromise, *not sacrifice.*

Sacrifice – the surrender or destruction of something prized or desirable for the sake of something considered as having a higher or more pressing claim.

Compromise – a settlement of differences by mutual concession, an agreement reached by the adjustment of conflicting or opposing claims, principles, etc., by reciprocal modification of demands.

See it this way—sacrifice and compromise will both become customs once implied or practiced in a relationship, therefore with time, our partner will expect them from us. The difference is that when compromise becomes a habit, it is a healthy ritual in a relationship, as the word itself requires a reciprocal action (involves two). In contrast, sacrifice is a one-person ritual and is not what a relationship is about. Besides, if the relationship turns more serious and both of you pronounce your marriage vows, you'll have plenty of options to practice your sacrificing skills. For now, let's leave that one on ice for a bit longer and focus on our compromise skills instead.

Back in chapter five, we talked about the elements of compatibility we look for in the opposite sex before dating: *attractiveness, intelligence, personality,* and to glue it all together—*chemistry.* As we move further into our relationship, we're going to have to learn to compromise in all areas, plus our five elements of love and healthy relationships and perhaps even more. This should not be cause for a headache as long as we have a 50/50 to 70/30 rule in mind. We'll call this our compromise boundary.

The 50/50 to 70/30 golden rule means you should never do something you 100% don't want to do and likewise never cross the boundaries of anything your partner or you don't want to do less than thirty-percent of the times. Meaning, your partner loves the opera (I do. Got a problem?), and you are not its biggest fan. Unless you despise it, going once or twice now and then won't cause any harm, but if your partner wants to go every weekend or every other weekend, you should be rightfully outspoken about this.

Whether it's a bar, an event, a movie, sex, and so on, it's not at all difficult to fall into unhealthy patterns to satisfy the person we're with since loving them can magnify our empathy for them. You should never give in to doing something you absolutely don't want to, and even the things you don't want to do should have a ratio because, calling it like it is, it's not fair for you to object to 90% of all the things your partner suggests either. Otherwise, you're not very compatible. Or, if you remember, "if everyone else is always the problem, perhaps the problem isn't everyone else," and maybe we should go back to the drawing board and work a bit more on ourselves.

I once heard something along the lines of "I know what I want and what I don't," and sure, that's all nice and dandy, Miss Prim, but being independent and knowing what you want shouldn't translate into being a pain in the a** and *always* getting what you want. There's a difference between an educated opinion and someone who *doesn't* take other opinions. Whether this negative attitude is adopted by your partner or you, it can quickly turn problematic, and the same is true for the opposite and always being complacent with what your partner wants. Before you know it, you'll either hear or find yourself saying "but we always _____ what you want to _____(fill in the blanks.)"

Compromise is all about not wanting to do something, but having a "meh, I'll give it a shot" mentality because at the end of the day, our partner's company is what counts, and we always have a good time together.

A good example would be a movie you want to see that's a genre that it's simply not his cup of tea. In that case, treat yourself; be happy with you. In a previous relationship, I remember wanting to watch a movie, and when I mentioned it to my partner, she wasn't feeling it, so I didn't think much about it. I bought myself a ticket, bought me some popcorn and a big ol' beverage, and treated myself to the Sunday matinee. Afterward, I went over to tell her all about it.

Don't emphasize so much on what you want or like and remember what we like doesn't necessarily have to be liked by our significant other. A relationship is about working together, not about having someone work for you. That's not from what the stuff of love is made.

Love in relationships is not unconditional. A lot of things are not necessarily reciprocal, and you know this book's stance on equality (yikes), but if there's anything that requires a bit of everything mentioned, it's compromise. Compromise is all about practicing our communication in the relationship and about doing things for each other, adding yet again one more element of building trust and taking us closer to love.

Trust

Trust is Addicting

Trust is as much about giving and receiving it as it is about being conscientious in recognizing the significance of having received it in the first place.

I often hear people in relationships say, "I give you my trust until you break it." If that's your open position, you're either dangerously naive or you have a kind heart more people could use. The bad news is they'll be doing exactly that—using it, and sadly, it won't be of much use to you. It's like lying down next to a hungry tiger and saying, "I trust you won't hurt me." Though I'm afraid the tiger is not nearly as dangerous as the animal you are facing—a human. Whatever the case may be, whether it is a hypocritical, wannabe-Mother-Theresa hard sell, or a genuine heart, that approach is merely dangerous.

Trust should not be given. It should be earned; it should be built, shaped, practiced, and molded by both parties in the relationship. Then repeat over and over again. Trust is not built overnight but is a continuing process. Then trust will naturally grow in the relationship.

I remember watching the movie Ransom in theaters with my parents when I was only eight, and one specific scene has stuck with me since. If you haven't seen the movie, may I say, spoiler alert? Tom (a multi-billionaire protagonist) and Kate's son has been kidnapped. Tom decides not to pay the ransom for his son and instead turns it into a two-million-dollar bounty on his son's kidnapper's head against the advice of the FBI. It is here that despite his wife not siding with his decision, the lead FBI agent confronts Kate about Tom's dangerous and controversial choice, asking her to convince Tom to withdraw the bounty and pay the kidnapper. Against all the odds, and just when it sounds reasonable for her to take the FBI detective's side, a moment of magic happens on screen when he asks her if she trusts her husband. Kate's response?

> *"Yeah. I believe in him. I stick with Tom. And we always manage to land on high ground."*

Trust is as much about not hiding your phone and being comfortable with your partner grabbing it as it is about you not looking through theirs when you've grabbed theirs. Or, as in the example above, trusting our partner through thick and thin, in life's most challenging situations. What you see is that you naturally build the bonds of trust so strong it makes you feel much more love for the person. Trust becomes addicting and a synonym for love in lasting, healthy relationships.

> *"Leeloo Dallas mul-ti-pass."*
> -   1997, The Fifth Element

Like Aristotle, the elements of love also have a fifth element. An element so powerful that all the elements combined would not suffice without it— arguably, perhaps consciousness would not even exist. The fifth element is the element of life—**sex**.

FOURTEEN

The Fifth Element—Sex

THE ELEMENT OF LIFE

> **"**
> *"What imperative does a grey box have to interact with another grey box? Can consciousness exist without interaction?"*
>
> *– Ex Machina, (2014)*

Sex. It is the basis for life to exist. We both enjoy it, and it is crucial for us to maintain a healthy relationship. But why is sex so important to us? Why do we love it so much and, conversely, sometimes say to ourselves, "Gosh, well, that sucked"? And can we learn something about it that will positively impact our relationships? I'm confident that we can. But do you agree? *Is* sex paramount to you or your relationship?

Whether you agree that it is, disagree, or are undecided on the matter, I think for the most part, we can all say we've experienced many different aspects of the spectrum—good sex, bad sex, great sex, moments when we didn't want sex, moments when you did but the sex-thwarting little brats wouldn't go to sleep, and others where our libido was so high that you felt like a donkey in the spring if you know what I mean. But if sex for whatever personal reason hasn't been a part of your life yet, continue to read as most likely one day it will.

It's not my first rodeo, but I'm not a porn star either, so I've had moments where I felt like a total bada**; others where I thought, *meh, I could have done better*; still others where it felt like I was trying to get a clam through a slot machine where I thought *what the eff, get it together, dude.*

But I've also had my fair share of experiences from the opposite sex as well. From moments the experience was so good I felt like a crazed heroin addict fearing having to go to rehab; to others where my male brain said *meh, it's just sex, after all;* to other moments that lacked so much chemistry it felt like a chemistry class discussion on how to build an atomic bomb, sadly without ever finding the explosion; to others where I wholeheartedly was thankful, I came out of it alive, wanting to hug my friends for being able to hold them one more day.

I'm sure you've had your fair share of memorable experiences as well, to say the least. So, amid this whole disarray, in what direction do we point the finger? Because, as you recall the many chats you've had with the girls about your bedroom complaints—how big he was; if he wasn't big enough; didn't last long enough; lasting *too* long; faking it; the creeping doubts of him not wanting it anymore; his inability to make you get *there;* him wanting it *all* the time that you're too exhausted; to how he was *stabbing* it, making one think he was trying to kill it, all while he thought he was a total renaissance man in bed; to learning, with a great shock, that men indeed fake it in bed too—you didn't exactly think you were a sex goddess reincarnate, either, right?

Having stated that, I hope we can peacefully conclude there's not a whole lot at which we can point the finger. Why point it, then? Let's scrap that and instead focus on learning a bit more about this potent and crucial element of life, why it affects us, how it affects us, why it's critical to our relationships in the first place, and what to do in the cases you're no longer all that interested in it, to ultimately find some consistency.

Because, after all, enjoying some good ol' sex with our partner, and enjoying it often, is essential to maintaining a healthy relationship, and it's ultimately what we all want.

Sexual "Flow"

After reading everything in this chapter thus far, why is it that both sexes experience such variance in the spectrum of sexual chemistry? Let me put it this way. Do you remember that potent cocktail of love at the beginning of the book in both of our breakup chapters that seemed to be harmfully radioactive?

Well, whether you thought sex was important or not before reading up to this point, what if I told that—from mutual attraction to the actual sex act, to experiencing an orgasm included terms such as evolution, sexual selection, "sire choice," classical conditioning, operant conditioning, GABA receptors, glutamate receptors, neocortex—correlated to things such as doing cocaine, heroin injections, and seizures in epilepsy? Hopefully, it should give us a slight idea as to why finding good sexual chemistry with a partner is not so easy as we expect. Our bodies are entirely different, and our purpose and processing may be as well.

On the contrary to common expectations, it would seem as if this complex level of finding a suitable mating partner was the designed intention all along. Despite commonalities in our mate selection and sexual experiences, women do have a more challenging time finding a significant other who will satisfy them. I'm sure that as a lady, your personal experience at times may seem as if this was an exclusive experience and a frustrating one at that. Nonetheless, you should rest assured that, on the contrary, it is entirely normal. Your body, from physical stimulation, to climatic stimulation, and the afterglow might be more naturally selective to properly gauge a partner for procreation and also to equip women for the *mother-infant dyad.*

So, what is "sexual flow"? First of all, the psychological term "flow" refers to one of the most enjoyable states the human mind can experience, which is being in the present, deeply immersed, focused, and engaged in an activity, while thoroughly enjoying it. Overall, a total of six factors have been recognized as distinguishing flow states (Csikszentmihalyi,1997): 1) intensely focused concentration, 2) merged action and awareness, 3) loss of self-consciousness, 4) personal effectiveness, 5) alterations of subjective time, and 6) intrinsic reward. In our sex lives, establishing flow means not much more than precisely that—finding chemistry, rhythm, and synchrony while being profoundly absorbed in the sexual experience and utterly enjoying it. Nevertheless, for different reasons, reaching a state of flow in sex is not that easy. But why is that?

Houston, We've Got A Problem

Aside from the strong cocktail of reasons mentioned in the paragraphs above, reaching a state of flow in sex is difficult due to the simple fact that we are simply anatomically different. While a man achieves an intense orgasm by a rhythmic, back-and-forth thrusting motion, making him feel a massaging sensation on his member (hence the "stabbing"), a woman reaches her climax/es by the stimulation of the clitoris. Basic Sex Ed 101, right? Yet, it would seem this isn't as obvious to many in our current society. Regardless, achieving a woman's sexual climax to where chemistry is established among the two partners through a state of flow is dependent on an array of factors that are very different for both sexes.

From attraction to climax, a man may require no more than an attractive body and rhythmic oscillations, perhaps because from what we learned in previous chapters, sexual selection is more straightforward for men. As you may recall in our men chapters, the male brain narrows its selection process to primarily two things: physical attraction for mating and personality traits deriving from different factors for the sake of being nurtured. A nurturing quality that may very well be subconsciously sought by men, to find the fittest woman for the mother-infant dyad.

On the other hand, the same process from the moment of attraction, sex, climax, a repeated climax to ultimately determine chemistry by women is much more complicated than men's, as it demands synchrony through at least three primary structures: 1) Increased combination of stimulation throughout the nervous system, 2) Enhanced mutual attention through stimuli perceived by our five senses: visual, auditory, olfactory, tactile, and gustatory, and 3) the outmost stimulating performance to elicit both a bodily nervous system response (somatic) and "reward system" (brain; dopamine release) response.

In much simpler words, a woman's nature demands a man's physical and fitness prowess in the attraction phase that correlates to the stamina required for the sexual performance ahead. To achieve a woman's orgasm during sex, a man will have to manage a continuing rhythm with correct posturing to stimulate the clitoris, which momentarily aligns with her natural preferred stimulation rhythm, while her body is also moving and gyrating to establish synchrony. Oh, and on top of that, once reaching synchronization, he must simultaneously use his five senses to stimulate hers to make her reach a climax, preferably multiple times ideally. The best way to imagine it is picturing *Interstellar*'s (Christopher Nolan, 2014) "it's necessary" docking scene in space while the other spaceship is plummeting to destruction. In other words—not easy (and purposely so.)

Therefore, it is because we are indeed two mechanisms with different bodily functions that finding sexual chemistry in a partner is not easy nor as common as we'd like to think, and not so much that they *suck*. Hence, why you're in control of the sexual act, you move in a motion that will help you gain *your* climax, *not* his, and additionally, he might fear you *breaking it*.

Quid Pro Quo

After learning of the complexity of finding and reaching sexual chemistry and climax, how could we ever solve such a riddle? *Sexual flow*. Sexual flow requires falling into a trance-like state, a lower level of consciousness, a higher perceptional state of feeling, and it takes two to achieve it. The problem is we automate our processes and lack understanding of them, creating assumptions or misinterpretations in consequence. The issue is there's much criticism, not much communication, and not enough letting go.

The solution? Allow me to happily introduce you to the element that requires all other elements of love, along with our foundation, to work in unison—sex. How? Via communication between partners, comprehension of the process mentioned above, and by letting go, effectively giving into our sexuality, while being aware of respect, honesty, compromise, and trust. The simplified version?

Comprehension meets → *Embraced sexual femininity = Sexual Flow*

A most common complaint women have in their relationships or dating lives about sex is that he's either not doing it right, or that he doesn't ever engage, requiring them to do all the work; both are right. Allow me to translate. In our first scenario, "not doing it right" basically translates to you not getting pleasure out of the situation or let alone reaching climax due to lack of chemistry, whether it's you, him, or both of you. In the second scenario, if he doesn't engage and allows you to do all the work, I'm sorry to break it to you, but this guy is merely looking to get his and has a complete disregard for your sexuality. Now, it's important to keep in mind that in both cases, it is normal if this happens occasionally. What we must avoid is becoming a developed pattern.

Nevertheless, you're not free of sin. Sex is still an act of two, and though you should indeed be the selector, there's much you can do before you cast the first stone. Indeed, if after you've played your part to increase your chances of enjoyable sexual success, then perhaps it is time to move on—and *not* into the relationship.

The truth is that the possible situations explaining why the two scenarios above occur are many. The scenarios can range from simple lack of compatibility to your partner not knowing what he's doing (due to inexperience, lack of knowledge, etc.), to you being rigid as a board or being non- participatory, to him not trying due to lack of regard, libido disparity, or in many other cases, a woman who generally has a hard time climaxing in general or the man climaxing too fast. These last two could have lucid explanations and solutions, but we'll get to them in due time. But as we've learned, focusing on the *whys* is focusing on the problem, and that's not exactly our philosophy here. Therefore, what can we do?

On your behalf, a couple of tweaks here and there could ensure your chances of enjoyable sexual success. On the man's side, there's still much more to be learned, and perhaps a sequel to guide us could be of help, but for now, let's focus on you, since teaching men is not, nor should it ever be your responsibility. The first thing you can do is be conscious of the fact that giving you pleasure is indeed more complicated than you probably thought. The second thing you can do is keep in mind the foundation— *communication* and the other four elements of love—*respect, honesty, compromise,* and *trust* as we'll need them in this text.

Thus, meet him in the middle (*compromise*) and guide him towards your pleasure by *communicating* with your facial, bodily, and subtle vocalized expressions. *Respect* teaches us to be supportive of each other's struggles if any should arise—or not rise. Be understanding that in any given situation, there's always more than meets the eye—and this is not exclusive to men.

After all, it is *your* pleasure you're seeking, so what insecurities could be intervening from being a bit more empoweringly selfish to obtain it? Because he's going to try to get his, so, how does being negatively critical help *you?* Being a little more open and selfless will help you be a bit more selfish and achieve your pleasure.

Just like it's not about teaching him, it's not about doing nothing, either. Practicing *honesty* with ourselves, our sex mate and the implications of sex asks us to be reasonable in all those three areas as well. Say that for some reason he's unable to get an erection that day. Most likely, it's on him, right? But what if it's not? Would you be able to hold the same judgment for a lady's lack of lubrication, or is that on him, too? How about him climaxing too fast? How about a lady who generally can't climax? Are both his responsibility?

A good tip is to praise instead of criticizing, let alone diminish. You don't have to give credit where it's not due, but guiding is more about where to go than where *not* to go. In essence, telling your partner what he's doing right is a hundred percent more effective than telling him what he's not.

Then, when we have remembered to practice our communication skills along with the other three elements of love and relationships, we must then give into a remaining component: *trust*. To trust is to do exactly that—let go. Not to let go carelessly or thoughtlessly, which is why we introspect implementing the elements of love. It's vital to understand we must trust and give in to our sexual nature; otherwise, what is there to enjoy? The ink on these chapters has not been for you to fear, nor for you to be defensive or naive, but to arm you with enough understanding that indeed, you have the courage to trust your actions and decision making.

Paint a Jackson Pollock

Personally, I've been lucky to make a living out of two professions that stand at different poles of the professional spectrum, art and business. In the world of business, a famous cliché is "it's not personal, its only business." Though you would think it's a simple concept to grasp, it gets rapidly thrown out the window in the first encounter with the more grotesque aspect of the business world. Therefore, usually, when having conversations about the two and trying to draw perspective into the world of business and art, I say, "Business is about turning all emotions off and making all decisions purely and strictly with rational thinking. It is asking yourself what would an advanced AI machine do? And whatever I imagine the answer to be, I do."

But with art, it is doing exactly the opposite while having only a guiding principle. It is turning off your brain and surrender to emotion. In those moments, even the sound of a waterfall can fill me with overwhelming sentiment and make me feel, even if for a moment, in love. *Ugh,* powerful stuff, but my point is that despite it being no more than an on-off switch to me if you were ever to see me in such states, it would be nearly as meeting two different people.

This off-and-on switch is what I'd like to teach you since I believe everyone could greatly benefit from learning this on-off switch mentality. You see, it seems to me as if people are afraid of feeling, fearful of entirely giving into emotion. The reason being that opening up our emotions opens up our vulnerabilities, which makes us more susceptible to pain and people taking advantage of us. But while shedding them keeps us safer from ill, it also keeps us safe from the great things in life, such as love.

But it doesn't have to be that way; two rights can coexist, remember? You can shed your vulnerabilities from the ill-intended while giving in all together to your emotions. Look at the following image to help you visualize the full spectrum of emotions vs. the safe range of emotions in which most people operate.

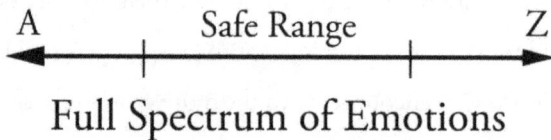

$$A \quad \xleftarrow{\quad\quad|\quad\quad \text{Safe Range} \quad\quad|\quad\quad} Z$$

Full Spectrum of Emotions

As stated, it works to keep you safe, but it also keeps you from enjoying the benefits of reaching A and Z and learning more about yourself. Instead, what I'd like you to try is this:

Conscientiousness > meets < Openness

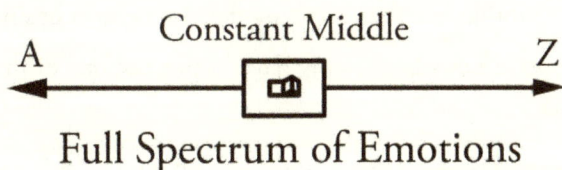

Constant Middle

A   Z

Full Spectrum of Emotions

Instead of keeping your emotions bottled in a permanent safe range, improve your level of conscientiousness by becoming more disciplined and *earning* your feelings through effort. Let them build up and then trust to let go entirely in those moments where you've earned it, such as the experience of sex. Every theory or philosophy is about our perspective. A problem is never the problem; our reaction to the problem is *always* the problem. Try to change your perspective. Don't see sex as more work after a long day's work, but look forward to the relief of stress in your orgasm.

As someone who is in love with art, I become naturally curious when I see someone work on something artistic or when they visit a gallery with me. In one specific case, my friend was painting a landscape while we were simply hanging out at her apartment during the weekend with some glasses (perhaps bottles) of wine. As she painted, I asked her what she felt, to which she answered, "Hmm, I don't know. Right now, about how to fix this."

I rephrased my question to draw out something more profound. "Sure, but what does painting make you feel? What emotions does it make you feel? What images does it make you see?" I was ready for the good stuff. My mind grabbed the popcorn, a soda, and grinned from ear to ear. My eyes peeled, waiting for an answer straight from the abyss of the subconscious.

"Hmm, I think mostly I feel relaxed."

There you have it, folks. Everyone, go home and expect no refunds on the popcorn (I still tease her about it.)

Exposed to such an emotional society, I thought being in touch with feelings was a less daunting task than applying the ideologies of business, but I was wrong. The good news is you only have to practice one of these two opposites—art.

The act of sex itself is an art. Sex is about letting go. It is a dance, not a competition; it doesn't have a plan or follows a specific order. Having sex is, in many ways, a fantasy, a dream, and our deepest desires. Think about sex as a split personality; so be whoever you want. There are no limits but those of your mind. Don't worry, rules don't belong in bed (or somewhere else), only elements as guiding principles, as even a great piece of art has a canvas. But within *every inch* of that canvas…paint a masterpiece, paint a Jackson Pollock.

Getting rid of insecurities and properly humbling the mind is critical to ultimately help you choose a partner that fulfills you sexually and intimately so that hopefully one day, you can form a family and make healthy babies if that's your choice. Because despite practice making perfect—and hey, don't get me wrong, I'm not one to miss a practice day—practice is ultimately to prepare for the big moments, to choose the right team that will get us to our intended destination. And the coach here is *evolution*, and the name of the team is *sexual selection*.

The Orgasm & Sexual Selection

In honor of Nathan from *Ex Machina* (2014), I'll direct you back to our original question in the script at the top of the chapter: *"What imperative does a grey box have to interact with another grey box? Can consciousness exist without interaction?"*

It's justifiably hard to think of something that feels so artistic to have such a banal-sounding explanation. But there's hardly anything banal at all.

On the contrary, it should only heighten the overall beauty of sex, life, and what we are as a living organism that forms part of a bigger purpose as a whole.

When we look at sex from an evolutionary standpoint, as in the scene from the movie quoted above, most people's first answer would be the same as Caleb's—an evolutionary need, a*nd here I thought it was just all fun and games, what do you know!* What an exciting proposition it is to think that sex and, perhaps, even love as a mere motivation for the rise of consciousness or maybe for its upkeep. And further, does that mean that language, sex, sexuality, love, dimorphism, gender, among other biological traits, are not necessarily to complement one another but to support the overall upkeep of consciousness in case one falls? Let me explain.

When we watch sci-fi movies or we talk about the future of AI, one of the controversial subjects that always comes up is the dangers associated with creating a conscious machine more powerful than the human mind; my question is, how? How would a more powerful conscious machine take over humanity? What would its motivation be without sex, without chemical reactions to spark emotion, art, feeling, love? Effectively, what imperative does a grey box have to interact with another grey box? I don't think it's possible. So, isn't it logical to believe an orgasm has a more substantial purpose than the one usually assumed by its society experiencing it?

The Mother-Infant Dyad

It will be easier for us to understand the connection between an orgasm and sexual selection if we dissect it into three easier-to-absorb components: the goal, the basics, and the purpose. Additionally, when explaining complex subjects, it's common for us to bump into the chicken-or-the-egg scenario, which complicates the explanation and, consequently, the understanding.

Luckily, I have a solution for this little pickle of ours in the form of an exercise better known as the "Three Whys."

The Three Whys is a little exercise to test someone's comprehension on any given subject, by asking three ensuing *whys* in consequence of the previous immediate answer. As previously stated, most people usually can't get past the first why. People often make the mistake of thinking of it as a stress test, but the exercise can be quite fun. Whether it's stressful or fun is dependent upon the resolution of the person engaging in it. It can be stressful if people don't practice thinking; otherwise, it can be quite amusing and fun. Try it out:

With a partner of your choice, pick a topic you like, something simple (perhaps a favorite hobby):

_____

1ˢᵗ Why: Why is that your specific hobby?

_____

After that, have your partner come up with the following two *whys* based on your answer.

All right, so as you come back to our lecture, we have engaged in the same exercise to get to the root of understanding the correlation between orgasms and sexual selection with the slight primary difference that we have expanded it to six *whys*.

Note:

The following exercise was done in a first-person perspective in the form of interaction, where the answers come from a completely dormant emotional state to help illustrate the point.

The goal:

(For heterosexual women) Learning and understanding to reach acceptance of the difference between men and women for you to achieve a better orgasm with your partner. Ideally, multiple orgasms.

1ˢᵗ Why: Why is having an orgasm with a sexual partner the goal?

Because it registers in my biology that if he's able to reach my paramount of sexual pleasure, he is the fittest and an alpha.

2ⁿᵈ Why: *Why does your biology register his performance to give you sexual pleasure as a sign that he is the fittest and an alpha?*

Because my biological structure and anatomy are designed for my climax and peak sexual pleasure not to be easily reached.

3ʳᵈ Why: *Why are your body and biology designed for your climax and peak sexual pleasure not to be easily reached?*

Because reaching my climax and utmost sexual pleasure takes physical fitness and achieving "sexual flow" by entering into a rhythmic trance-like state requires stamina, diligence, attentiveness, and focused perception while engaging all of his senses to achieve it. Therefore, this tells me he's the overall fittest because his stamina says he's healthy, his diligence says he is most likely to persevere in carrying out an action, his attentiveness tells me he has a caring nature, and his ability to focus and his perception show me he's intelligent.

4ᵗʰ Why: *Why are you choosing those characteristics in the first place?*

Because on a primal level, my body anticipates me carrying human life, even though medical issues may prevent me from doing so, or I simply may not want to. Life only chooses to improve as it moves forward. It must filter the weak from the strong for the species to thrive through the eons of time. Thus, only the fittest and strongest can survive that rigorous test of time.

5ᵗʰ Why: *Why does your reproductive need make you choose the strongest to thrive through the rigorous test of time?*

Because aside from the overall reproductive need of natural selection for my offspring to carry life forward, I have a reproductive *purpose,* and that purpose needs a companion to help me and protect me as I'll be at the most vulnerable period of my life.

6ᵗʰ Why: *What is your reproductive purpose?*

My reproductive purpose is the mother-infant dyad.

The mother-infant dyad refers to the mother-child relationship during the pregnancy and the first nine months of the infant's life, together totaling eighteen months. Though like anything in science, studies can be debated, the data backing the importance of this relationship is nearly overwhelming. Even from a biological stance, the rise of consciousness is due to child-rearing, nurturing, and bonding that humans experience being the social beings we are.

The proposition that the female orgasm serves as a filtering function in mate selection is known as mate-choice hypotheses. There are two recognized versions of these hypotheses: the sire-choice hypothesis and the pair bond hypothesis. The sire-choice theory states the female orgasm influences the choice of mate selection to pass on better genes. In contrast, the pair bond hypothesis states the orgasm influences mate choice by focusing on paternal investment and higher care.

As to the detailed literature of these hypotheses, you can find endless articles on the subject, and you can rest assured there will be some good recommendations at the end of this book with the citations' pages. The topic itself is too ample for us to dissect in one chapter and would ultimately deviate from our point. The point of us having a surface understanding of what sexual interaction, pleasure, and an orgasm implicate is for you to better appreciate the beauty within the complexity of your nature. Hopefully, heightening the value of choosing sexual partners with a better scope, thus filtering out the jerks that have no other intention but using you.

As evidenced by introspection and meticulous research, it is often seen that women occasionally fall head over heels for these "fitter" men in bed, often at the expense of their emotional stability. Besides, all science talk aside, what do you think about the matter? What is your take after introspecting? Be honest with yourself. As we said before, there's no one watching. It's a strict relationship between you and the ink in this book. So are you guilty as charged, or do you perhaps have a girlfriend who is guilty as charged? This challenge is not to contradict what all the pages of this chapter are trying to illustrate. On the contrary, it's to give you the perspective that despite sex being an essential factor when choosing a partner and keeping a healthy relationship, it is not the *only* factor in selecting that ideal someone. Sure, practice your hand at contortionism and have your fun, but engage consciousness and be truthful and realistic in summarizing the content you've absorbed for you to make better decisions in finding your next special someone.

It is important to remember the overall components and strategies that give body to the anatomy of love because when it's all said and done, sex for women and men is just not the same. Sex without the attachment of your personality and your affection is simply a delight, whereas the same sexual pleasure for a lady may very well have emotional implications of attachment.

A Man's Value Thinking Structure

To give you a male's perspective, let me let you in on a little secret. If it's been said again and again that a men's primary intention is sex, don't you think they'd be observant of your choosing?

Your body and sexual intimacy are a *prize* to a man, make no mistake about it. Do me a favor, and please try to turn off your female brain for a moment and dive into this notion with me briefly—really entertain this as it's vital. In the mind of a man, the work he puts in to obtain that reward will be the value he attaches to the compensation once he's received it.

In other words, the initial importance a man attaches to a woman will depend on the effort required to have sex with her, and once they do, it will depend on how good it was. Then, the blindfold will come off from his sex-driven thinking to see all of her worth.

Coincidentally the work they put in to obtain sex, your body—the reward—is **your** choosing; hence **you** choose the value.

With that information in mind, ponder about the following. A man sees another man in a relationship who is a good-looking fellow; he has his life in order and seems to be a good guy putting work into his relationship. Yet for one reason or another, his partner breaks up with him, and moments later, she's sleeping with another guy who put in only a fraction of the work than her ex, often just hours of effort. Tell me—tell men, why should men be the first guy (in the relationship) and **not** the second guy? Who is obtaining the *same* sexual goal by doing significantly less?

Upon reading this, you might feel judged or objectified, and I can't begin to tell you how gravely I hope it doesn't as that's not remotely the intention. Maybe it doesn't apply to you; perhaps it applies to someone you know, or maybe you *can* relate. This text is by no means telling you what to do. It is merely telling you how men approach and think in a woman-man dynamic, that's all.

What I'm trying to illustrate to you is that you have the power and control over this dynamic. From a man's point of view, it's tough indeed to understand a woman's grievance when the next guy comes along and wants nothing but sex when lack of chivalry has been previously celebrated. You are creating single personal experiences, training a man at a time. So let's do this, let's meet in the middle. It's not fair either for you to limit your sexual appetite because of all the men in the world; it's just not. But it's neither reasonable nor logical either when a woman grievances about a male who just didn't appreciate her despite not gauging her intimacy.

The same can be applied the other way around. It's as if I complained about not being able to express emotionally from the first moments in dating to a woman, and when I encountered women open to my complimentary side, I used them for only sex. It's fair to say, you'd think I'm hypocritical, wouldn't you? Therefore we don't get both; we have to meet in the middle to make it work. *All* experiences do not encapsulate *everyone*. There are bad apples and great women out there as there are men. Go out and get the quality man you deserve and have fun while you're at it, but don't fill a void of love with your body, as you are so much deserving of more.

Sex & Relationships

Aside from choosing a partner, being selective, orgasms, and more, why is sex so important in our relationships? And why is it often so complicated to maintain it? Well, just like the numbing effects sex has on the mind, opening the door to love at the beginning of the relationship—keeping the flame lit by having great and constant sex during the relationship is the key to not exiting the door of love. Otherwise, we'll find ourselves carrying the breakup chapters as our friendly companions.

Now we all know about those first months into a relationship where lust seems to be in the air, and you can hardly keep your hands off each other. Now, this stage of love, bonding, and feelings of euphoria making you want to check your temperature every time you see them, is known as **limerence**. Limerence is said to last anywhere from six to twenty-four months into the relationship. There on, the average couple has sex about 51 times per year, being that you visit the family one week out of the year and all, you know?

That means that couples, on average, have sex about once a week. But even then, that average is taken from some couples reporting having sex twice a week, to others having sex once a week and others once a month—yikes!

In one particular relationship, our sex was great. It was often as once per day to sometimes more than once in a given day (shame!—GOT[6]). It stayed like that for a good three years, despite the stats on couples. Then as if it was a matter of fate, it dropped as quickly as a mortgage fund in '08. So what happened in my relationship and throughout so many relationships alike?

We become complacent, plain, and simple. We get demanding jobs that model modern slavery; we get older, and we're tired all the darn time. And as much as we hate to admit it, doing the same thing over and over again with the same person who, as a matter of fact—*hush, hush, get closer*— happens to look a lot different than when we first started dating, well, for lack of a better word, is simply—*hard.* But I hope you can remember a couple of chapters back saying something along the lines of, "no one said it

---

[6] Based on the bestselling book series by George R.R. Martin and created by David Benioff and D.B. Weiss. "Summers span decades. Winters can last a lifetime. And the struggle for the Iron Throne begins."— https://www.hbo.com/game-of-thrones

had to be easy." The truth is that nothing worth having is easy, and an ideal, long-lasting relationship is not either.

I know that's what happened in my situation, at least. Because as much as I'd like to blame my partner at the time, the job, whatever—excuses are plentiful—it doesn't take away from the fact that I simply stopped trying. I gained weight, sex was accessible, and I was still climaxing, so why try? If I could answer the old me, I would probably say, "Well, to keep your relationship, *dummy!*"

From the time I was a kid, my mother has been very competitive. She has always stayed slim, jogged, worked out, ate healthily, and looked as pretty as possible. Back then, in my youth, I remember my mom going to school to get her bachelor's and my dad working a regular office job from Monday to Friday. Before he got home, my mom would always get ready, shower, dress nice, put on makeup, and so forth. One day I asked her why she did that every time Dad came home from work, and her answer has stuck with me since.

"Because I'm not about to become the stay-at-home mom. You see, son, the reality is he's a professional, and he sees professional, attractive women regularly. But, I know he's a man, and how men think, and also how women are to a man of his position. Then after being in that environment all day after a long day's work, they come home to their wives who sometimes haven't even showered. We let go of ourselves, gain weight, and when taking care of the children, the men sometimes even start calling them 'Mom' right along with the kids. Then they wonder how they found themselves in divorce court, cheated on or left for another woman. No sir, that's not going to be me. I'm finishing my career, staying healthy, and staying competitive."

Do you see where I get my competitive realism? Though it may sound a little biased to use my parents as an example, the truth is they might not be the ideal marriage, and I don't necessarily agree with many of the traditional aspects of their married life, either. But it is not my business, and they're still married to this day. It's also quite neat hearing how they talk to each other. It's quite a delight being on the phone with my mother and hearing my dad joking in the background, so I guess to an extent, you could say she was right.

The crude reality is that whether it's a new car, a movie that you've watched, a rigorous goal of finishing your first book, or sex in our relationships, they are not the same when they start to when a significant amount of time has passed. Let's be honest, would you want to watch the same movie over and over again? Except for *Forrest Gump*, but that's not even fair; that's more like a cinematic experience. Back to our point—the truth is that it's unlikely you would want to do the same thing repeatedly.

Now, those examples are emphasizing on materialistic and abstract concepts. Relationships take a harder toll on monotony. The truth is we are living beings who age, gain weight, and wrinkle; our hormones lower, and add repetition—well, yes, it's not easy.

So I guess we should pack our bags and call it quits, or accept our new relationship life as it is, or start scheduling appointments with sex therapists and stacking the meds, right? Well, of course not silly goose, there's so much we can do and accomplish from holding ourselves accountable and putting in the work.

Now, on the other side of the coin, that's not to say that loss of sex drive, lowered libido, and orgasmic anhedonia are just a figment of your imagination and that there are no actual conditions or situations that require the help of professionals—of course not. There are plenty of problems that would require treatment and much else, and you should seek the advice of professionals in those situations. But in the vast majority of cases, you and I have plenty of work to put in before scheduling a visit to our local sex therapist any time soon.

What Can We Do?

First of all, as you read, it's about staying competitive and putting in the hard work. It's a simple realization that while a pet's love and our parent's love might be unconditional, our romantic love is *not!* Just like we had to work to *obtain it*, we have to work to *maintain it*, and the world out there does not have to and will not accept us just because of who we are. Therefore, the very least we can do is maintain the image we brought into the relationship when we first met. This notion applies as much for men as it does for women. Truth be told, while on my jogs at the park, I'm always surprised to see couples with kids, and in a lot of cases, the mama looks indeed like a hot *quesalupa* while the man in the relationship makes me wonder if *he* gave birth to the kids.

One

Stop embracing the dad-bods and start embracing healthy-looking bods. Exercise naturally does three immediate things for us in our sex lives. For one, it keeps us looking nice and tasty for our significant other.

Secondly, it increases testosterone levels in both women and men, which maintains, and in many instances even increases, our libido, keeping us hot and ready for our loved one. It also increments estrogen levels, which are essential for the ladies if they want to maintain and prolong those years of putting babies in the oven of life.

Also, as you may remember from our breakup chapters, it makes us produce our own opioid-like drug right from our spine to lower those pains and rusty bones, better known as endorphins. Okay, so maybe that was more like four immediate benefits instead of three. So what? The more, the merrier.

Two

The all too popular world of porn and the self-indulgent act of masturbation. *Oh boy, where is he going with this?*

While everything with limits can be harmless, the reality is that the stats tell us a whole different story and hint that we might be overdoing it in this department. At the very beginning of this book, it stated that love was "the third-most searched word worldwide (the first two not being so benign)." Well, would you like to take a wild guess as to what the first- and second-place winners are? If you haven't guessed, since you have been reading for twelve chapters, those two are "sex" and "porn," to effectively *not* have sex, and *effectively* watch porn and masturbate.

Women get the upper hand on this one as they are beating men in the porn department and are not prone to watch as much porn, watch as graphic porn, and sometimes not watch porn at all, as they have switched over to reading erotica. Like men, what women *are* doing wrong is indulging too much and too often, but we'll get to that in a minute. You may want to let your male companion read this next part.

The world of porn is affecting men, and the research is overwhelming. You see, let me tell you a little something about the mind. Though it is a beautiful machine, and the structure of the eye is no less than marvelous, overall, it is still not incredibly great at telling reality from fiction. Sure, you're a conscious being, and you can consciously tell the difference between your own a** and elbow, but your brain processing, not so much.

In the breakup chapters, I briefly mentioned how turning on the TV provided you with a tiny bit of the stimulus you need from socializing and that your brain isn't great at telling the difference. Well, what I was referring to is this—a heterosexual male brain is designed to elicit a reaction from the bodily features of a woman, therefore causing an erection and affecting the testes' sixty-four-day sperm production cycle and consequently affecting testosterone levels. So how does this affect you, stud?

Since your brain is getting constant stimulation from porn, it affects your overall fertility as your sperm production lowers. As sperm production decreases, testosterone does as well, which dramatically reduces your libido. Also, by stimulating your brain from porn and elevating the graphic content of the porn you watch, you accustom your mind to unrealistic expectations. Studies suggest the level of a man's expectation can be somewhat measured by the number of tabs you open as you pick your crème of the crop porn videos. Then, when you involve yourself in the real thing, those expectations cannot be met, and expecting your lady to meet them is not only unrealistic but unfair. Lastly, as a natural and obvious consequence, porn causes us to masturbate often, which also affects us. But this is now a subject of two.

Therefore, how does constant masturbation affect both women and men? *Wait, is he talking to us?* Yes, ladies, I am, to you and your vibrating, 1000 rpm, competition-murdering friend. Like nearly everything else we have discussed in this book, it's all about meeting in the middle.

Masturbation is a natural act that keeps us healthy, stimulated, and allows us to get to know ourselves better, which is crucial for a healthy sex life. But doing this far too often has consequences of its own in that it complicates our sex lives by creating the opposite effect in women and men. In men, it can lead to premature ejaculation, non-prolonged erection, or not as firm erections. Which if you don't know how that would affect your sex life, well, may the force be with you.

On the other hand, overstimulation for women can cause you not to be able to climax at all, as if it wasn't as complicated as it is already. Thus, it can nearly destroy your sex life and your ability to gauge the competence of men, as if *that* wasn't as difficult as it is either. Therefore, self-indulge, get to know yourself, but don't overdo it—it's overkill.

As important as it is to get to know thyself, it is incredibly vital for us to get to know each other. Perhaps try merging the best of both worlds in bed. You may want to try using your sex toy as you're having sex with your partner. At the very least, by achieving your orgasm in this manner, your brain will relate the positive experience with your partner, the more you do it (classical conditioning.) This could naturally increase your desire to have sex with your partner until you can eventually achieve your climax with your partner alone.

It's also important not to ignore our verbal communication skills. It's normal to experience a bit of shyness when talking about the subject with our partner outside the bedroom. After all, chatting requires the great majority of our thinking brain to be engaged, known as the neocortex (the wrinkly part that makes up 76% of the brain), and specifically for our brain to turn on the decision-making part of it known as the prefrontal cortex.

Ironically, that decision making part of our brain called the prefrontal cortex gets "shut down" when we're having sex, to allow us to get into that trance-like state we discussed, hence, why you can barely utter a sentence when indulging in the act.

Because of the reasons above, it's recommended we practice our communication skills outside the bedroom and talk about what makes us tick, what doesn't, and more. Think of it as a *rehearsal* to a scene vs. the *actual scene* = *Verbal* communication *outside* the bedroom vs. *bodily* communication *in* the bedroom. Stopping a scene to go over your lines wouldn't exactly be the best time, would it? Likewise, stopping the artful act of sex that requires the five senses, trance-like state to achieve a state of flow, etcetera is not the best time to express your verbal skills, either.

That said, it is also important to experience new things with your partner and step out of your comfort zones to see what else is out there in the world of sex. Perhaps read a book about some of the oldest practices, such as the Kama Sutra. When sex diminishes in relationships, it's more an issue *getting* to the sex than *having* sex. This dilemma is often because the poor lad has the same three moves for you as he did at the beginning (and, hey, we do try very hard, so don't be too harsh. Who said you were—okay, I don't have to say it [introspection]).

All joking aside, it could also be that you've felt your libido lower as of lately. Thus, setting specific times and days for you to *enjoy* some playtime might be necessary if you have kids, but if it's not required, then try to refrain from doing this by all means. Making sex systematic will most likely detriment your love for one another—Remember that sex is an art, and there are no rules to art.

There could be a vast amount of perfectly sound reasons why, as a woman, you don't feel like doing it lately with your partner. From taking birth-control pills that shoot estrogen and progesterone (most commonly) into your system daily. To not feeling as physically attracted to your partner lately, to exhaustion due to routine, or experiencing a lowered libido. Women sometimes experience decreased libido after thirty-five. The one thing you shouldn't do is punish yourself. You're not less womanly or less beautiful of a design because you don't feel like having sex.

One of the best things you can do is talk about it. And while it's okay to talk about it with your significant other, talking to ladies with a similar experience by joining a group or venting with some girlfriends might be a great solution. Pamela Joy, the founder of "Down To There," focuses on raising awareness to "Reclaiming Female Sexual Desire" and states that most women keep these issues to themselves and are as surprised and relieved as she was when they find out they're not alone. Know that you're not alone; there are many things you can do to turn your relationship sex-life around, but keeping it bottled up inside is hardly one of them.

Therefore, it is crucial to experience the early stages of love, during the honeymoon stage and set realistic expectations for what's to come and build sexual standards for your relationship.

Also, remember to be proactive. If you're the lady in the relationship, explore the sexual realms of his imagination. If you're the man, remember she's not an object for your sexual leisure. Use your words, your seduction techniques, and show her there's plenty of love and life in you still and that indeed she has seen nothing yet. Caress her, kiss her, and use your soothing words and romantic tone to conquer the queen of your empire time and time again.

There are many options to experience, maybe close to endless, and you by no means have to try them all and even less meet your partner's deviant expectations or become a porn goddess. Nor does he have to become lothario god in bed. Remember: compromise, not sacrifice. Therefore, experience away. Sex doesn't have to be a daunting job, but it does take work.

Remembering something as simple as the love for one another can go a long way when attempting to surprise and support each other. The human mind is an extraordinary complex and powerful machine. We'd be wise not to underestimate its ability to automate when we lack action, which will quickly form vicious cycles that kill relationships. Let me explain, *just like our brain elicits our actions, we can elicit our actions to program our brain.* A small example of this would be:

I feel sluggish ➔ I can't smile ➔ I look in the mirror ➔ I feel sad

Instead, what you could do is:

I force my smile ➔ I look in the mirror ➔ I laugh ➔ I smile naturally ➔ I feel better.

In our sex life, the process is more complicated, and it takes more time, but it works the same way. Look at the following illustration to visualize it better.

Brain says:

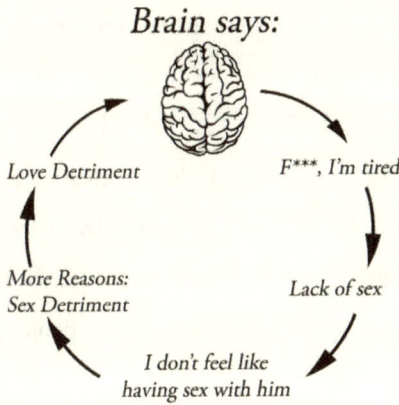

Love Detriment

F***, I'm tired

More Reasons:
Sex Detriment

Lack of sex

I don't feel like
having sex with him

Fortunately, we can reverse this process with merely with two ingredients—love and proactive effort.

Brain says:

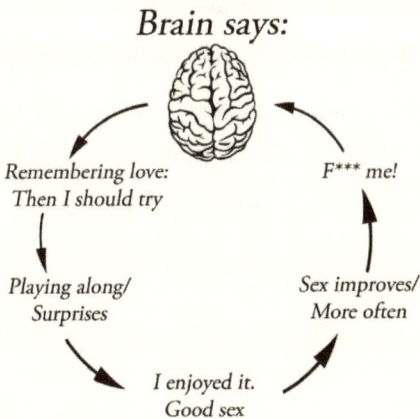

Remembering love:
Then I should try

F*** me!

Playing along/
Surprises

Sex improves/
More often

I enjoyed it.
Good sex

Just like our brain elicits our actions, we can elicit our actions to program our brain. What that phrase means is that when we lack motivation for something, making ourselves do it regardless of the catalyst with time can program our brain to engage in the action naturally and finally find a natural reason.

Try to remember to essential ingredients to this recipe—surprises, and playing along. Surprises imply no more than the definition of the word. It means you could surprise your partner as he comes home as you're wearing some naughty lingerie. It means that perhaps he can surprise you to a fancy dinner looking sharp as you come home while a new dress lies on the bed.

On the other hand, playing along means that just as your partner put forth effort trying to surprise you, you should put effort into playing along and act surprised whether it did or not. This point obviously applies to both women and men. It shouldn't be all that hard, should it? We already fake it at parties, and social gatherings, when meeting new people, at work, and even when the minions ask our thoughts on their new drawing. And we smile and say how wonderful it is despite not making the top from the bottom, don't we? So what moment could be more important than actually putting forth some effort towards the act of kindness the person with whom we decided to spend our life is doing for us?

As a business owner, I can tell you pockets of hard work make the reward so much better down the line. So do the same, try it out, and have plenty of fun with your sex life as you do, and to you and that special lucky someone, I wish you the best of luck in the many long, happy, and healthy years to come.

PART FOUR

An End's Beginning

I tend to be social, and though I am charming
my emotional logic lacks.
So my id hides up in the attic
hushing the rhyme that follows the _ _ _ _.
If your wit finds the answer, then please go and help her,
As the key lies in this riddle,
And it also construes a person's wrath.

FIFTEEN

A Love Story or a Story About Love

WITHOUT A NAME

"

Every new beginning comes from some other beginning's end

— *Lucius Annaeus Seneca*

" *I want to see you today Darya and thank you for your kindness, but I sincerely don't feel like waking. I'm in bed without much energy.*

With one eyelid half-open, he types on the screen that shines a painfully bright light at him.

But this gem of a woman won't have any of it. *Shut up and get up, it's a new day. Let's go. Meet me at Memorial, please.*

Beep. Another incoming message pops up in H.'s notifications bar. He's grateful for the concern of his friends, but he barely has enough energy to wake up, to think, let alone get up. Still, he opens the incoming message. It's from Skeet.

You okay, bro?

Not really. H.'s response subtly extends the length of an appeal.

Are you in pain?

I feel... I don't even know, beyond depressed. My mind is failing me. I feel demoralized and tired.

Well, express it, that's the first step.

I don't feel like talking about it.

Dude, I get it. You're super smart, and I love you for it, but this is not the time to stay in a dark room and sob while you over-analyze this s***. You need your boys, and you need to vent.

All right, thank you. I will. I'm going to see Darya in a bit. Can we meet later?

Sure, brother.

***

Nine months earlier

July 2019

A Love Story

"Absence sharpens love, but presence strengthens it."

It had been two years since H.'s last relationship, and the week before, he'd quit his job to pursue working on his businesses full time—a goal he'd envisioned and planned for a long time. He had met beautiful women since his last relationship and had a fling here and there, of course, but hadn't found anyone with whom to spend the rest of his life.

He was adamant about not making the mistakes he'd made in the past and rushing into a new relationship. But despite feeling a bit lonely as of lately, he'd convinced himself a relationship could wait, as this new business venture that felt like jumping off a cliff with no parachute just to see if he could fly demanded his full attention. Perhaps it was a sign, he thought, since the last two women he'd genuinely been interested in hadn't materialized into something more for some reason or another. Besides, it'd been five days since he'd last received a message from another lady he was interested in but hadn't met. Because despite not being religious or believing in a superior power, he admitted there was something more. *It must be for a good reason. Dating can wait. For now, I must submerge myself in my work,* he said to himself.

But just when a new journey awaited him, another new journey began.

Beep.

Hi, I'm sorry I didn't respond sooner. It's been a busy week for me, and **I was dating**, *so I don't know where things are going now.*

He couldn't help but feel excited as he saw *Her* message.

Oh, no worries. I'm not in a hurry, so whenever you feel comfortable, I'd still very much love to meet you.

She must have met some disappointments—I'll be patient, he thought.

You make me smile. :)

And he did too. There was a unique hint of affection in *Her* reply that at the time, he couldn't really make out why he liked so much. Now he'd tell you that below the layers of his stern expression and a firm resolution was a man longing for a woman's warm touch. He was a man who no longer craved the friction of a massaging aesthetic frame after sundown, but the intimacy of a lover's bare anatomy lying next to his in the waking of the light.

That Was Smooth

A couple of days after that exchange of messages, they met at Waterwall Park. He'd been there countless times and was taken by its delightful scenery every single one of them. Thus, he thought it'd be a great place to meet, aside from making her feel safe at a pleasant public place.

He'd be lying if he told you he wasn't a bit nervous—and so would any other man. Yet, he put on a show of confidence with every stride as he approached her.

"Hi!" *She* greeted.

"Hello. What a pleasure to finally meet you. I have to tell you—and I mean this in the best of ways—I'm glad you're not as tall as I thought you were."

"What?" She laughed.

He grinned from ear to ear. "Yes, you look tall in your pictures, and to be honest with you, I was a bit intimidated by that... and you do look like Anne Hathaway, by the way."

"I think I have one of those familiar faces that looks a lot like someone people know because I get it *all* the time. I've always been told I look like someone they know. I don't know, maybe it's the symmetry of my face."

"Oh, really? Maybe..."

They began to stroll around the park as they continued to talk.

"Have you ever gotten that?" she asked.

"Actually, no, I don't think I have. At least not often enough that I can recall. Opposite of you, I have a very peculiar square head"—they both laughed again—"that I think prevents me from looking like anyone at all."

Every step was accompanied by the soothing symphony of sounds from the human-made waterfall as it crashed at the bottom steps, and passersby admired it looking up.

The sun blessed them with milder heat this evening. The weather granted clear skies, which in unison painted a view with a palette of a bright green turf, trees with darker shades of green, a cerulean sky, and a bright white stream that complemented the view from the opposite side of the staggering stature of Williams Tower standing in front of them.

"Had you ever been to this park?"

"No, I haven't. I've heard of it but never been here."

"Why don't we take a seat on this bench over here?"

"Sure."

Once they sat, he gazed directly at her and couldn't resist smiling. "You *do know* I can't wait to talk to you about psychology, right? I didn't study it, but as you know, I love psychology."

She returned his smile. She was a clinical psychologist doing her residency in Houston. "All right, let the probing begin."

Their pleasant chat carried on, and the clock quickly consumed the minutes. Before they knew it, night came, and darkness bathed the sky. He wanted more of her company, and she didn't show signs of wanting to leave him yet, and so he proposed they go to a nearby bar, to which she agreed.

As the night progressed, they touched on an array of subjects, some casual, some intellectual (the ones that excited him the most). He'd always had an insatiable wonder for knowledge and rarely ever got the chance to have meaningful conversations about them—even among educated people. It was fair to say he found most people he knew to simply be uninterested in learning, just in knowing enough to get by. It was an opportunity of which he took full advantage that night, and despite perhaps making her head spin, he was simply enjoying himself. Besides, she didn't seem to mind. It was also fair to say H. could talk anyone's ear off, granted he had the proper subject.

But what love story could ever be called such without a nutty, memorable moment? Perhaps he'd gotten so carried away by his own chatter he entered a trance-like state—but no one knew for sure, and it is still a mystery to this day.

Just as he became conscious of his yapping and came to a halt, he came up with the marvelous idea of stating the obvious (though he didn't think it was in that moment). "There's no need for us to come up with something to say. We can have comfortable silences." (Legend has it he still bashed his head against the wall whenever he remembered that moment.)

His date answered with a proper and straightforward, "Okay."

But stating the obvious wasn't enough for our witty protagonist; no, he thought to spice things up with a dash of irony as well.

"I just think people usually want to fill the air with something to say, despite not having much to say about a subject."

She smiled, nodded, and stared.

Our hopeless romantic protagonist was somewhat of an introvert—*a detail that should be kept handy.* Yet, despite H. not being the most sociable, one could say he fended off pretty well in social environments. But as he was about to find out, not so well on this particular evening.

As the silence filled the atmosphere after his glaring remark, for a reason he still would not be able to explain, he became a self-conscious, nervous wreck. But, the show was far from over, and it seemed H.'s date was enjoying the spectacle, because she smiled and stared, and stared some more.

It was that fine evening H. would learn of a condition he hadn't heard of before—*hyperhidrosis.* Meaning sweating when nervous and then sweating some more when one becomes aware of the sweating, and then some more, with the dreadful anticipation of hopefully not sweating more (which he does). Picturing Brendan Fraser in *Bedazzled* when he's a famous basketball player sweating uncontrollably in front of the reporter will paint a neat image of what H. was suffering at the moment.

"I'm enjoying your vulnerability right now."

Are you now? How nice. I can't say I share the feeling, he thought. After digging such depths of obviousness, he thought he might as well add some more. *What the heck.*

"Yeah, I'm going to have to excuse myself," H. said. "This is ridiculous. I became self-conscious after stating the obvious. I'm going to the restroom. I'll be right back."

Once in the men's room, he quickly dried his face with a paper towel and coached the guy in the mirror.

Dude, what the f***! Get it together. You haven't even kissed her. Are you really going to ruin such a smooth night with this? No, you're not! You got this.

He took a deep breath and headed back out without hesitation.

"Hey, there, thanks for the help, staring at me while I was embarrassed out of my mind." He joked, but his words rang with confidence again as he sat composed. They both laughed for a while at his predicament, and it wouldn't be the last they would.

As the night grew older, the music seemed louder, and their smiles more frequent. Their laughter was still young, and their conversation flowed with indulgence. Before the bar closed, he took care of the tab after *She* offered to pay half, and they walked together to their respective cars. He leaned with his back to his as she stood in front of him, preparing to say goodnight, at last, comfortably allowing the silence to take over the ambiance while they exchanged amorous glares. He slid a hand around her waist and tugged her close, savoring her gaze a moment longer before claiming her lips with his.

Their lips parted slowly. *She* found his gaze with hers and said to him, *"That was smooth."* A moment as he'd like to remember.

The upcoming morning, the light that squeezed in between the blinds woke him, acting as his alarm. He felt a little discombobulated, as expected after a late night of alcohol and fun. He smiled, remembering the night only a couple of hours past. He turned to the other side of the bed to find her curved figure still resting peacefully. It had been a celibate night, as she had requested. It had been merely a night of new romance where she allowed his limbs to become lost, wandering through the curved hills of a silk-smooth surface trying to find their way back home. At long last—*the intimacy of a woman's body lying next to his in the waking of the light.*

Losing Track of Time

TIME MOVES FAST WHEN YOU'RE HAVING FUN

Leo's Birthday

Only a couple of days away from his birthday, they celebrated that previous weekend with some of his friends at an Asian restaurant she had picked out for the occasion. It was Saturday, July 27th, when his fondness for Bill Bryson began in the form of a gift from her. The night came and went. Spicy margaritas became the night's trend. As the hours consumed, he reached for the pretty lady's hand by his side, and he celebrated in delight under the umbrella of the stars.

A Fiesta Under the Phoenix Sun

The kidnapping that never happened became a running joke among her friends. During the first week they started dating, H. had an upcoming trip to El Paso to visit his parents, and though it was a bit too soon in their relationship, he invited her to come with him. She thanked him and told him she would get back to him about it. In the upcoming weeks, as they spent more and more time together, she inquired about a heavy noise coming from the trunk every time he turned. His answer? "A body." (Put your phone down, it was his bike rack.)

"I knew it. My friends warned me about it. They said you invited me to El Paso to cross me over the border and sell my body parts."

"I mean, it's a profitable market. Besides, white girl organs sell for double the price. But I guess I can't anymore—bummer."

"Son of a *B!* I knew it!" (Again, relax. She was joking.) But perhaps it was a bit too early in the relationship, after all, so she took a rain check. Dot the *I*'s and cross the *T*'s, people say. Instead, when it was a bit more appropriate, they planned a trip together to Phoenix.

Their itinerary included a hike, during which he fell, bled, and twisted his ankle (shouting out her name as he was falling)—only to find her moments after to the sound of, "Oh sugar! What happened to you?"

They stayed at a beautiful, cozy Airbnb in which *Men in Black*-sized cockroaches came out of the sink, forcing him to show his "manliness" and move the fridge, stove, and kitchenware before killing it while she laughed hysterically and nonstop.

The afternoon turned into the evening, which turned into a night bathed in alcohol. The girlfriend he knew clocked out, and an alter ego appeared and decided to entertain her hand under the table in his lap while they had a conversation with their new bar acquaintances. They effectively blacked out and consequently woke up in a haze the next morning, finding food all over the counter, a bottle of Tajin, and a glass filled with red wine—and *milk* (no one was injured.)

It was an eventful adventure, to say the least, under the scorching heat of the Phoenix sun. An evening they finalized with the pronunciation of *girlfriend* and *boyfriend.*

In a short amount of time, they lived a lot. Slow living started moving fast, and as time does, it moved on. The clock watched as it went along its monotonous path. The minutes consumed the hours, the hours consumed the days, and the days took the form of weeks, and then weeks morphed into months that took the heat of summer and transformed it into the cold breath of winter. With it came and went the patient strolls through galleries and museums, the late nights that proved regretful headaches in the morning, in which their bodies reunited in a waltz of lovemaking under the welcoming sunlight. And as life is no fairy tale, then came what seemed a relationship-ending fight.

"If we can't resolve this, perhaps it's better we break up and…"

Before he could continue, she interrupted him and pronounced for the first time, "H.—*I'm in love with you!*"

And the symphony played on.

He thought she knew more about living, and he knew more about life. They continued busy living, one trip after another—Jacksonville, Dallas glamping, Mexico, Denver. With the trips came meeting parents, talks of having kids—to the extent of naming them, moving in together, Sunday fun-days, a chubby little lady feline, and a lovely tuxedo cat who was also an as***** with a "can't touch me" anthem.

Their life together became visits to the zoo where elephants danced to the rhythm of Shakira's music. Assembling puzzles as he did a piece at a time as she completed five. It became finding bunnies, meeting friends at breweries, cheering her on as she ran her marathon, watching pretty sunsets, battling spiders in the car, and wearing ugly Christmas sweaters strolling through the Lights in the Heights. In other moments it was taking in the pretty River Oak lights during Christmas time. Then came the trips to the mountains, the ocean, and no-singing-karaoke dancing nights, another fall in Denver as he slipped on ice. Endless outings, the pizza, spicy drinks, the ever-present wine, and constant food-tasting that turned into extra pounds.

With it came the pictures, captured moments, endless hours of shopping, smiles, and blurred laughs. And just as quickly as time passed while they were having fun, just when it seemed things for them were going fine, and H. couldn't help feeling happy to the point of joyful, proudly saying "she's all mine"— only for nine months after—breaking up, and waking to a sun that no longer shined. Slowly thinking, hating, he should have seen the signs.

A Story About Love

LIFE HAS AN ODD SENSE OF HUMOR

Hi, I'm sorry I didn't respond sooner. It's been a busy week for me, and **I'm dating someone,** *so I don't know where things are going.*

Oh, no worries. I'm not in a hurry, so whenever you feel comfortable, I'd still very much love to meet you.

It was a reply from H. that might have never been.

You make me smile. :)

Life has an odd sense of humor. Who could have ever thought a single message could change an entire timeline of events? Otherwise, if he'd read what the message said the first time, he would have never responded in such a nonchalant manner. *Perhaps it would have been the story that never was.*

The Twenty-Sevens

Friday, March 27, 2020

Despite the blinding mental fog, the numbing pain, a vanished ego, and shattered self-esteem, he managed to make it out to the park with his friend

Darya. Primarily because he understood the importance of getting out and also a valuable friendship, so he didn't want to object to her kindness, being that he was in short supply of it at that moment. Yet, despite the calamity he'd just endured, it seemed a male's biology never stopped working, as he found her stunning. A sight he thought wise to show the world of social media as a sign of defiance against his current state. She obliged for H. taking a picture of her and even knotted her shirt into a crop top. Perhaps, even though he didn't yet know it, in that very moment, something had snapped, lit, changed—*sparked.*

"Everything seemed to be going fine, H… How did you find out? Were you expecting it?"

Was I? What a great question, he thought. "Perhaps I missed the signs. But no, I wasn't expecting it at all. If anything, I thought we were working things out."

"Ugh. And to think I liked her when I met her. I don't like her now." Darya paused as her frown dissolved. "Be honest, though. Hadn't you told me about her friend?"

He thought about her challenge for a moment, allowing the breeze to clear his thoughts. "It wasn't that, but I guess you're right. Ever since the beginning…"

That Was Smooth?

July 12, 2019

"How about you come over to my place?" H. asked ***** after their first date. "We can have a drink and relax by the balcony."

"Sure, we can do that," she replied. "The only thing is, I don't want to build expectations for sex tonight."

"Yeah, that's completely fine."

Once they were back at his apartment, he decided to make himself a drink, while she opted for water instead. With a pool view, sitting on some nice rocking chairs, their conversation continued.

"So, if you don't mind me asking, how is it you're single? You're beautiful, smart, educated. Have you just met a bunch of douche bags or what?"

She chuckled.

"No, not really. I've been with wonderful men. I guess it just hasn't worked out."

Hmm. Then why hasn't it?

"I hear you. Was it a long relationship that just didn't materialize?"

He was wary of why a woman with all those qualities happened to be single—basing it purely on measures of attraction—attractiveness, intelligence, and personality—as far as he could tell from the first two, she was a good catch.

"Well, we dated for two years. Things weren't working out, and then he just moved out. I was hurt, so I cheated on him. So we separated."

He didn't love the cavalier manner of her answer, but at least for that moment, his brain wasn't thinking for him.

"Oh well, I'm sorry to hear that. I've been in longer failed relationships myself. The last one being four years before we broke up, so don't feel that bad." They both smiled.

She tried to add some humor herself. "Yeah, I've dated emotionally distant men, so when things start to fall apart, that's usually when I go kiss someone else." She laughed.

But he forced his. But he didn't want to hear more. He got up from his chair that was a couple of feet away from hers. For the second time that night, he kissed her.

She looked at him and said, "That was smooth." Perhaps with time, he convinced himself it had been a different scenario.

The acquainting of their bodies carried on for a while more. He felt like an explorer in a treasure hunt, and with every discovery, he yearned for more. But as she kept her resolve, he tamed his burning desire. And yet she stayed, something he found peculiar since he hadn't personally experienced someone spending the night when sex wasn't involved. Having always been enthralled by psychology, he wondered if this moment had anything to do with attachment. An idea he dismissed by enjoying the scene.

"You still have your clothes on," She said. "Hmm. Do you sleep with clothes on? Why didn't you take them off?"

"Well, we didn't have sex, so taking my clothes off felt awkward because I didn't want to impose."

As he continued to court her, every time they met, she vetoed his advances, and every time his lust for her grew. But one day, after a weekend's drinks at a nearby bar, she suggested it with the casualness of a Monday morning meeting.

"I couldn't wait to be with you, you know?" he said to her humorously, as *She* smiled.

"All right. Are you ready for us to do it every three weeks now?"

She laughed.

She spent a lot more time over at his place. He'd be lying saying he minded all that much. Notwithstanding the chemistry, the signs, he ignored the inner weakness that whispered to him how afraid he was of the idea of being alone. Some say that a man's work is confirmed by the woman by his side. But they also say people often try to amend the present with the guilt of broken pieces from their past. An idea he dismissed from his sight, as he wasn't about to allow himself to lose the image of all she represented in his life. He was blind by his success-driven craze—or perhaps, *self-doubt.*

As the weeks passed, he grew to enjoy her extroversion more than he cared to admit, since he was on the polar opposite end of the spectrum, naturally introverted. He loved with burning passion every aspect of his life—what he did for a living, the privilege of working from home, his habits, such as learning about different subjects in science. There was something about the human mind, the power of knowledge, and the infinite question of *why are we here?* That completely captivated him. He loved introspection and the feeling of growth, like sculpting his body, and even just thinking. But he understood people's views on these matters and that they found it— *boring.* Even his dating profile stated so: *"I'm by the book, what you would call boring. I like…"* Which was what she said she liked about him, his genuineness.

Sometimes he thought she knew the city better than him, despite him living there seven years longer than her. Everything they did together was new for him, hence why he enjoyed the experience of giving in so much, changing his firmness from the past despite sometimes feeling anxious.

Saturday, July 27, 2019

Birthdays are usually an exciting day for most people. But he was nervous. There was a lingering irony to the thought of someone thoroughly enjoying the depths of the mind and being genuinely interested in knowing people. Yet, H. didn't find social gatherings to be the best medium for that purpose. He would say he saw social gatherings as a bit—*disingenuous.* His closest confidants described him as somewhat of a straightforward personality with good intentions that sometimes caused people to feel uneasy. He, on the other hand, would say people understood him as much as he understood them; in his mind, he simply didn't comprehend why no one ever spoke their mind.

His personality rarely allowed him to celebrate birthdays. It'd be more likely for him to celebrate accolades before celebrating his birthday. But this one was a bit different in his eyes. For one, he got to show off to his close friends who he hadn't seen a while how well he'd done in the last couple of months. Not only had he just opened a new business and left his job to focus on both of his businesses full time, but he also got to show off his new, hot, smart girlfriend to his friends. He had been carefully selective about bringing someone into his life in the past two years he'd been single, after all. Thus, he thought, *what moment could be more warranted than this?* Besides, it was the first time H. and *She* hung out with anyone else aside from each other, and the company of her natural charm put him at ease. Thus, he would make sure to look like a complete stud for the occasion as he prided himself in his hygiene and fashion.

But above all else, he wanted to impress her. On one previous occasion, she'd made a complimentary comment about her friends, who he hadn't yet met, and he wanted to show her he had quality friends as well. However, before getting the evening underway, some of his friends had already called, saying they couldn't make it due to the kids, work, or else. So right off the bat, things didn't go exactly as he planned—nor would they the rest of the night.

Once at the restaurant, the typical chitchat took the main stage for the night: *how have y'all been? How are the kids? How's work? How did you both meet?* In total, there were seven of them, three couples and another friend—Skeet—who hadn't taken a date that night. As was typical, conversations shifted in different directions at the table. So, as H.'s date chatted with his dateless friend so he wouldn't feel left out, H. talked with some of his longtime friends Anita and Ross, who sat to his right at the end of the table. They mainly caught up about how things with their one-year-old daughter and their current pregnancy were going.

"Oh gosh, I can't wait for him to come out, he's so big now," Anita said with a big smile.

H. and the other couple chuckled, accompanying Anita and Ross. As the apparent joy remained on the expressions at the table, H. reached for his date's hand to remind her he hadn't forgotten her.

"But we're enjoying it while we can because we know short sleep hours are coming."

"Your firstborn trained you well, I see, huh?" H. said. "Well, you look great, Anita," Oddly, his date hadn't yet taken his extended hand, which now rested awkwardly on her lap. "I can't wait for the new baby. I'm excited for you both."

"Thank you. We'll let you know for sure when our baby is born so you can come meet him."

"But of course, count me in. You know, I'll be there."

"Which reminds us; Anita and I have to talk to you about something." They both smiled with grins of suspenseful excitement.

"Surely, we'll get together soon."

The food tasted great, the alcohol flowed down thirsty bellies, and the smiles were in abundance. As the other couple joined in the conversation, H. turned his attention back to Skeet and *Her.*

"...and the second one is a tuxedo cat. James and Gordita."

Skeet laughed. "I really like those names."

"Oh, you're telling him about your cats?" H. interjected.

"Yeah, she is. I was just telling *Her* how I used to have cats myself and how much I miss them."

"So, do you have any pets now?" She asked Skeet.

Still, *She* hadn't reached for H.'s hand, he felt her farther away than usual, and she hadn't even looked at him as he joined the conversation. Something felt off about the interaction. He thought her a bit too smiley and friendly. *Was he perhaps feeling jealous of his new girlfriend?* But he waived off the uninvited thought. He had never been the jealous type, but at that moment, the whole scene made him feel uneasy. Yet, he kept his smile for the rest of the night, afraid someone would notice him.

As the evening persevered, it was all an act—he enjoyed none of it. His hands were clammy, and his cheek muscles started to sore as he counted down the minutes in his head, waiting for the night to end. *-Beep-* Despite personally disapproving of the use of phones at social gatherings, he attended to the vibration emanating from his pants' pocket. He justified himself with the thought that perhaps checking his phone would distract him from the sour moment. It was a message from Skeet. Just as quickly as he read the message, he put the phone away as he picked his head back up to meet the liveliness at the table with a smile.

With bellies full and a kindled buzz, Anita and Ross invited H. and his date over to a nearby bar while the other couple called it a night. As they walked out of the restaurant, H. tapped Skeet on the shoulder discreetly hinting for them to have a brief moment.

"Hey babe, I just opened the car. Would you mind giving me a brief moment with Skeet; I won't take long." His voice was calm, and his performance didn't waver.

"Is everything okay?" She asked with an intrigued look.

"Yeah, Skeet just wants to tell me something real quick. I think it's something about his date; no worries, I'll be in the car in a minute." But he lied.

Skeet and H. made out words through their teeth as they stood in the middle of the parking lot. With the SUV running, and the driver door far-flung, Ross awaited. Few friends could be as faithful; Ross waited as a guard on duty for no other reason than making sure H. was—*straight*—as he would say. But as he did, he managed to overhear H.'s part of the conversation, who stood closer to where Ross was.

"…yeah, no worries at all brother, I appreciate you for letting me know. Just a misunderstanding, that's all."

"You sure?" Skeet asked.

"Yeah, of course. We'll talk more when I come back from El Paso."

"Alright, bet!"

It seemed a bit random, but after those brief moments, they both smiled and parted ways.

Yet, before he could take a step, Ross approached him as well and got in close to tell him something. Again, it was no more than a moment before they headed back to their car.

"Alright, I'll see you there at the bar!" Ross said aloud. Before turning around towards his car, where his date still waited, H. took a deep breath as he produced a slight smile on his face once again.

At the new bar, they all enjoyed some spicy margaritas that seemed to be perforated from the bottom as the drinks expired quickly. The conversations here flowed with more ease with every passing minute, and the buzz slowed H's thoughts. A serenity that would not stay without interruption.

"So is Skeet coming?" She would ask for the second time that night. To which H now responded with a tired and apparent-forced smile.

"No, I don't think he's making it tonight."

"Why couldn't he make it?" *She* probed.

"I don't know, *Her.*"

After the night had ended, once home, he reflected in silence. Sitting with one leg hanging over the armrest, nursing a scotch, he allowed the full force of gravity to take over his body. He stared blankly, thinking, growing upset as he reread Skeet's message.

Hey brother, I'm not making it out to the bar. Hit me up real quick.

He redirected his gaze to an empty space in the room, letting his thoughts drift off as his jaw line tightened. Now a man a year older, he sat there, remembering the words of Skeet and Ross, along with his lies to cover up his feeling of dullness, with every thought, now feeling—*embarrassed, mad, puzzled.*

"Hey, bro, we have to be brief, but I wanted to bring something to your attention because you're my boy. Out of respect for you, I'm not going to the bar. I'm not one to tell you about your business, but since I saw you notice in there, I thought I'd let you know your girl made me very uncomfortable." Skeet's words echoed inside the walls of his mind, nearly blending in with those of Ross'.

"Hey, what's up with your date man, everything all right? I'm super oblivious, but Anita was telling all about it, so I wanted to check on you. We can call it a night if you want. You tell me."

It was supposed to be my day. I was to show them how well I'd done. How can I do that from my position now? He thought. With a knot on his throat, he felt—*stupid*—for the first time. But not for the last.

Finnegan's Wake

After that event, he felt confused about the decision to move forward or move on. He wondered if continuing dating *Her* could prove painful. At the same time, he asked himself if leaving her could mean missing out on a good match. Due to his analytical ways, he always based his decisions by weighing pros and cons, and would often even write them down—and he did.

The way he saw it in his mind was, on the one hand, he had her pros, and on the other, her cons, but as a second determinator, he had life.

Life

Pros	Cons
Safe, steady dating – may lead to a relationship.	Being alone or out in the dating game again.
Could coincidence be a sign? Why now? After two dates hadn't materialized and a week after quitting his job for his business.	He needed to focus on work. Was it a sign of life testing his resolve?
It could mean love	It could mean pain

Her

Pros	Cons
Good qualities: attractive, smart, career path	His birthday incident – a big no-no, and it was embarrassing.
They had a good time and liked her social choices – museum, art galleries, holes-in-the-wall bars	Little comments he didn't like – staying over, her joke about inconstant sex, how easily she expressed cheating in the past.

She had seen past two little unlucky coincidences – nervous sweating moment and car not starting due to a dead battery. It could mean she was sweet.

She continued accepting money from her dying ex. Could mean a lack of remorse or empathy.

Yet, even after jotting his thoughts down with pen and paper, he felt as confused as before. *What to do?* He wondered. So, he seduced himself with the idea that *time* would give him the answer. He thought either the right or bad traits would outweigh the other in the upcoming weeks. If her negative characteristics became more pronounced, he would end things before letting his guard down. All the same, if her positive traits won, he would then let his guard down and let himself *experience* fully—*win-win.*

Besides, in the last three months, two dates hadn't worked out, a gorgeous Colombian and another ravishing Latina. *What the heck, third time's the charm, right?* He convinced his optimistic mirror reflection.

What he didn't count on was that when you let time pass and give in to the experience, that's pretty much all it takes for the potent formula of falling in love. And when you fall in love, indeed—*you fall.*

***

With the passing weeks, the Sunday-fun-days became regular leisure, and so did the outings and late nights. Thus came wine tastings, museums, and romantic settings under cozy, red lights. But at the same time, also went the subtle, quiet aggressions that she called expressions—the sly dubious comments and with it, the fights.

It stayed like that for a while. With weeks, days, and hours getting blended as one, the experiences also mixed many smiles, as they traveled for miles from the Phoenix' scorching sun to the outskirts of Dallas, coming together as one. Slowly, the adventures, the moments, and a slick, hidden omen, formed little pockets of fun.

At first, as their relationship started patiently pacing, they soon began racing, and slow-witted walking turned to a run. Soon after, H. noticed himself falling in love. He thought his thinking was perhaps juvenile. But his feelings objected to the sick-driven notion thereof. He dismissed it as an absurd and vile motion stemmed from doubtful, lonely-fearing emotion— and not a calling of love.

Then when he felt that things were going just fine, her behavior again began changing to seeking attention and creating tension, which he saw as a sign. But she claimed it was not her intention, so he asked probing questions, and despite always making great conversation, her answers were short, distant, and trite. The whole situation gave him a gut-wrenching feeling while seeking answers to make up his mind. But soon, he would find that it's not easy to decide on what's right when a mind-playing player leaves answers open for interpretation as her grad education has narrowed her sight on leaving the men-painstaking memories behind. All while the man she was breaking gave her his undying love and affection, but to that, she was blind.

In the end, their situation seemed like a blend, a mixture of a love-hating growing frustration in both of them, which had no end in sight. Or maybe not quite. So because of the confusion, the ups, the downs, the love, and the relationship-ending fright that turned into constant threatening, due to their relationship bind, he decided to avoid conflict, the friction, and ultimately a fight by avoiding verbal confrontation and instead he would write.

H. had never identified to be indecisive, but quite the contrary. This feeling was new and foreign for him and something he wanted to express to his girlfriend. But whenever he tried having a similar conversation in the past, it had always turned into an unproductive discussion. She would tell him that he was too forward and reminded her of someone with whom she never had the best relationship, nor gotten along all that well. In his mind, she was very defensive whenever he wanted to address anything about the relationship.

Not only did he genuinely love her and cared about her, but her personality incredibly intrigued him as he would later admit to a friend—*he had never met anyone with such an interesting character.* Hence why he decided that writing her a letter, dissecting his thoughts would be the best avenue to take. H. believed in working things out with those you love, and he also believed in hard work, since that's the way of thinking that had gotten him thus far in his life. Besides, they had dated for months now; he had acquainted with her friends whom he'd grown fond of and made great memories together. They had also recently come back from spending eight days visiting her family—indeed a lovely bunch. He was also ecstatic about her meeting his mom, who was visiting in a few days for the Lights in The Heights.

Once he finished his letter, he thought he would kindly propose it to her, but to his demise, before he could begin to explain what it was, the pure thought angered her, claiming it was *everything he saw wrong in her* before reading it—she did not. Yet again, being upset wouldn't be enough, and just days before his mother arrived, right then and there she would abruptly—but ever so calmly—break things off.

It seemed there was always a demoralizing power play with her. First H.'s birthday, then bringing over his belongings another time after a full day of fun after he simply mentioned he didn't like her comment about her friend's boyfriend——only to tell him "…I'm in love with you" for the first time after he had agreed to part ways. One day she loved him, another he was as disposable as a trash bag. Nothing in the middle could exist with her. It was utter love or utter neglect, and this time it seemed it was the sheer humiliation of a broken promise of Lights in the Heights to his mom. A lovely lady, the light of his life, who confessed to them she was a little nervous about being in a full room of people from a different culture and all who were doctors on top of that. When the time came, though, they were all sweet, welcoming, and great as H. knew they would and his mom felt welcomed and liked them all very much.

As he took his letter back, he folded it slowly with the composure of a dog with his tail in between his legs, shamefully burying it in his pocket. The thought formed *I just wanted to talk some things out* as a knot the size of a fist produced in his throat, feeling impotent, stupid towards her actions one more time. She knew what his mom's visit meant to him since she knew of his story—an only child, with an abused past, to a struggling single mother, now visiting her son to meet his girlfriend and a promise of spending a fun weekend with his *new friends*—or so he thought—ironically, now at the disposal and hand of a psychologist. He began making words withholding his need to cry.

"I apologize if I offended you, I just wanted to talk. I can improve; we can make things work out…"

Yet he still didn't know it, never had his mind been *fucked* with so much.

Two weeks before March 27ᵗʰ

"I think we should break up," she said.

To his surprise, he felt a breeze of peace after kissing her on the forehead. He was proud of how he acted. He walked towards his car with a slight smile, his shoulders thrown back and his head high. He felt chivalrous, like a man who accepted whatever wasn't on his hands, and like Pilate, he washed them of sin. But not everybody had the habit of walking away with clean hands.

The next day he felt surprisingly free. He treated himself to a deli burger, dedicated time to work, and watched a movie in bed before calling it a night. Yet, without question, he had her in mind. The next morning was similar. He washed the dishes, cleaned the kitchen, picked up the apartment, and as painful as it was, he removed their pictures from the fridge and undid the puzzle on the table, feeling as if with every piece, he was taking a piece from his heart.

Then he received a text message from her.

Hi. I was wondering if you had time this morning for us to talk and for me to pick up my things?

He replied, *Sure thing, I just did some chores. Come on over.*

Once she arrived, the mood felt different. H. smiled, felt confident, but was determined to be welcoming. They greeted each other with the formality of those who haven't seen each other in months. While he brewed a fresh pot of coffee, she walked around the apartment, noticing their missing pictures, the puzzle, and let out a sigh. Once she'd done a lap around, she sat on the couch.

"Would you like some coffee?" H. offered.

"Um . . . how about some wine?"

They both let out a genuine laugh.

"Ah, what the hell. Sure, why not?"

What transpired next was pleasantly unexpected. Like two birds who'd just met, they flirted, laughed, smiled, filled the room with affection, disregarding the apparent subject entirely, and for that moment, his heart beat with hope.

After an hour or so of this pleasant act, he said, "Hey, have you had anything to eat?" As she searched for an answer, he continued, "Why don't we go grab something to eat and have a drink?"

"Yeah, sure."

They would go out together to a neat Mexican restaurant-bar with a lovely patio. Like all meaningful conversation, theirs ranged from an array of topics, jumping from one to another as they consumed margaritas and a variety of mixed drinks. Mostly they joked, laughed, and even talked about their zodiac signs, pointing to some of the hints of astrology that read: "Leo is in polarity with the fixed air sign Aquarius. Ruled by the sun, Leo likes to express, conquer, and perform, where Aquarius (classically ruled by Saturn) likes to test, judge and set boundaries. When in the opposite sign of his preferred home, Saturn is said to be in detriment or disadvantage in the bold sign of the Lion. Imagine the stern planetary god of time, trials, and discipline, thrust into a world of drama, pride, and passion (astrology.com.)"

Still, after reading and discussing their astrological alignment, he got up boldly, walked around the table, got within a couple of inches from her, and said, "So, do I have my girlfriend back?"

And just like that, reminiscent of their first encounter, he laid his lips on hers, sharing a patient, affectionate kiss. She smiled as they separated, and for a moment more, he thought to claim once again what was the love of his life.

"And *now* I have to go to the men's room. I'll be right back."

The night sailed smoothly on, and once again, he felt *happy.*

But real life seemed to be doomed of happy fairy tale endings. The ship was dull, safe, and "steady as she goes"—or *overboard*, it goes. What started as some drinks of liquid courage seemed to turn into a liquid toxin, which she spilled as *She* saw fit as if from nowhere.

"But we haven't even talked about the sex," *She* said. "We don't have sex."

"I know," he replied. "I've brought it up before."

"I don't need sex. I can go without it for weeks."

"Is it me?" As if in slow motion, every syllable he pronounced told him what a stupid question it was. Welcome to the self-crucifixion of martyrdom incarnate—*may the wrath of a hundred lashes begin.*

"I just don't want to have sex with you!" she said. "I don't know. It's your smell, your shape." *My smell?* He had a hard time accepting her words, as they were words he heard for the first time. But that's important to you, and I don't know I want to be with you."

"Why are you talking like that? What's the need? Is that something you'd like to hear from me?"

It was a question that forever went unanswered.

Frustration Attraction

As the week progressed, he remained stern and composed. He accepted *life's resolve* despite longing to be with *Her.* Yet, they kept in touch—then just two nights later, they watched the movie he had bought the night they broke up. A night during which she forgot her cell phone case—a coincidence that hadn't occurred before. After she texted him in the morning inquiring about her case, she went over to his house to pick it up. They had some coffee, sat down, chatted for about an hour, and then parted ways.

A couple of days after, he saw the bike he'd just bought her forming dust on the balcony, and to satisfy his melancholy, he invited her on a bike ride, since they'd never been on one before. Once there, all of the beauty and green of Bayou Park could not overshadow the blues and awkwardness of the whole interaction. It was distant, cold, and too proper, and he was beginning to feel drained of love. So he kept to himself, accepting life's resolve.

The next day, *She* contacted him again to let him know she was going to archive their pictures for her sanity and well-being as looking at them was too painful. Her message left him wondering why she needed to contact him to tell him about that, *a reason he would later find out.* He expressed his thoughts—"I think that sucks, but okay,"—but accepted that which he could not control. He took it with a simple "Okay" and welcomed life's resolve.

As the weekend passed, her memories haunted him. So that first Tuesday, he invited *Her* for coffee and a walk. Mainly because he wanted to see her because he'd be lying saying he didn't love her. But also, because of the strange effects of ill love, the impact of giving someone a taste of you to plunge them into a feeling of ecstasy, and later sending them into total despair when they face the adversity of boundaries and physical distancing, which heightens their romantic passion. Like a child's sense of their parents' abandonment can lead to rage, stemming from the frustration of not being with them, heightening the craving and the longing for their affection. Despite official terms and special practices, it was something one would expect a clinical psychologist with a Ph.D. in neuroscience—as she was— to know.

They strolled along, talking about life's mundane occurrences, which after a centurial pandemic that had just begun weren't all that mundane after all. As they walked, she got a call. She took a step back, smiled, and turned away from H. to send a text.

Well, that wasn't obvious at all, he thought. He knew too well what the provocation was, but what right did he have to air a grievance? Besides, his pride would not allow him to say anything.

He had been posting more regularly on his social media since he was shifting his focus to work on himself and making it public put pressure on himself to get things done. The next day she commented on one of his stories, and once he responded, she said nothing. She wouldn't answer all day, which he found unusual. But she eventually responded close to midnight, which he found odd since she always went to bed early. The human mind, at its essential core, is a pattern-recognizing machine, so he noticed this discrepancy. He found it difficult to believe that such simplicities of the mind escaped her and was all due to the innocent mishaps of sheer coincidence. He had acted with maturity and patience, an attitude that few, in his opinion, would. Exhausted by whatever senseless game *She* was playing, he decided he wasn't going to chastise himself anymore.

The Bottomless Pit of Self-Loathing

March 26th

Emotionally drained, he made up his mind to confront her once and for all. What he hadn't expected was how much she expected *him*. He told himself that civility was most important, and against all, he was not going to tarnish his worth. Otherwise, for what had the last more than two years of introspection and growth been? Apparently, to be used as a carpet for those whose past demons still haunted them. He was full of anxiety, feeling like his heart was pumping nitric acid. His moist hands left imprints on the steering wheel. He knew all too well about the best practices to keep his mind at ease, inhaling deeply and exhaling slowly. He repeated the act again and again as he practiced posing his question since his mind was racing. There was no such thing as a sixth sense in science, but it was still remarkable how people tended always to know the answer to the questions they were asking ahead of time. They just couldn't help but see it with their own eyes, and H. was no exception.

Once at her place, he put the coffees down. The formalities were out of the way without wasting time, but not before he noticed a bottle of American whiskey resting next to the wine he'd left there.

He sat across from her, took a deep breath, and said, "I want to ask you a question."

She nodded. "Okay."

"Without the slightest intention of conflict—are you seeing someone else?"

Her answer couldn't have come with less hesitation. "Yes. I just started. Why?"

It didn't matter that he knew the answer. He, like anyone else, was hoping to hear something different. Her response translated to a sudden shock in his chest, sending a surge of pain to the innermost part of his core. Most of all, it was her proud demeanor, her poise, that filled him with disbelief his mind could not decipher at that moment.

"Why?" He wanted to scream and cry and plead what a thousand words could not suffice, but nothing else aside from a simple *why* came out.

"Does it matter?"

Over fifty facial muscles control a human's expression. At that moment, as she sat across from him, every one of her facial muscles was stretched back. Her posture, her dead gaze pulled at the corner of her eyes, her tight, poorly concealed smirk, the tone of her voice, the brevity of her replies, the rectitude absent of all warmth that had existed only moments ago. The woman he knew…was gone. Where was that sociable person everyone else saw? Where was that soft voice that inspired innocence that everyone heard?

He forced himself back to the present, and though his consciousness shouted for him to leave, to let go, his curiosity sparked by the strength of affliction. Surprisingly, his composure was not yet lost.

"Of course it does. You were in my bed just days ago. I'm not arguing with you, but I want to know why. I want to hear you say something more."

The truth was, he wanted no more than a peaceful closure and her kindness one last time. *What could be so difficult about that?* He thought. The questions crowded at a barrier that couldn't hold much longer. But it didn't matter; she would hear none of it. The empathy of humans extended to the length of their need. Her need at this moment was the dopamine craving feeling of power, that once *She* tasted, only demanded more.

"I'm not required to give you an explanation. You and I are not together anymore."

Like a tourist trapped overseas, sitting tight as the overseer used every tool to remind you how you're no longer welcome—your privileges have been lost as your passport expired. Thanks for flying, don't come back to visit.

"Honestly? Get the fuck out of here with that," H. said. "Why are you even acting like this? What have I done? I've been nothing but a good boyfriend to you!" He thought perhaps a fresh reminder of his relationship accolades would inspire a different response from her. "I loved you. I've been nothing but respectful to you."

As he listened to himself talk, he felt he started resembling a babbling baboon, quickly turning into a muffled intercom speaking to deaf ears either way.

She maintained her poise—concrete, distant, and cold. "Yes, you have."

Confused—pained—depressed—demoralized? He wasn't entirely sure how he felt, but most likely all of the above and at the same time. The experience began to take its toll and completely overwhelm him. He felt sick. He didn't know whether to throw up, burst into tears, or shout. His fortitude officially subdued, he forced out words in the tone of a neglected child trapped inside the body of a man.

"I know it sounds naive, but if anything, I thought our relationship was going well. Why were you still seeing me?"

"I was talking to you to help you." The answer rang so incredulously it was hard for him not to gag. She got up with grotesque determination and an expression of disgust, now repudiating who just days ago she'd said she loved. "And now it's time for you to leave. Please go."

She didn't even care to look at him as she commanded him. The abuse was intolerable, and someone else came out of him.

"No!" He snatched her phone from where it rested on the table. His fingerprint still worked. Perhaps digging wasn't wise, but it was too late. He was in a trance, possessed as he scrolled. She demanded her phone back. The levels of toxicity poisoned the room with every passing minute, but the urge was now in command. She demanded it once again, and upon seeing his lack of response, she proceeded to act two.

She began to cry, collapsing to her bottom and adopting the look of a lost child. He couldn't believe his eyes and thought it worthy of an award, but he'd heard of this act since childhood and had seen many faces of this act since. He remembered something his mother had once told him: "Listen, always be mindful of the things you hear, and do not be easily persuaded by anyone, men or women. Use your brain and always think critically. Unfortunately, we women often cry and play the victim to get our way, and once we have, we smile inside as we wipe our tears."

As he read what he shouldn't have, his pain morphed into something without a name. His mind blurred, and the image of the woman he'd given so much to was forever erased. He saw texts that said *I miss you—I wish I were still by your side—* explicit sexual requests, specific attire requests. They had exchanged work schedules, and as he confirmed, the new man had even given her the bottle of whiskey that was next to his bottle of wine. But he did not need to worry, as she blatantly admitted to all of it, she set his mind at "ease" by assuring him it had been using protection. All of these things resembled a woman he did not know; they were things he had never said himself out of his respect for her, and her denial of such advances as she mocked him saying, "it's as if you still want to live a porn fantasy." Yet, he had lived plenty of fantasies, just none with her, a woman he loved, and with who he just wanted to share their intimacy. He read of things she had never done with him.

"What a clown I've been," he cried the moment he realized all of this had happened between them in no more than four days.

She wiped her tears away as if someone had said "cut" and said her warmest words to him yet. "I was proud of myself for not cheating on you."

The onslaught went on for a while longer. H. was the senseless cattle walking to his execution. Before he left, he made one more request, an appeal that stemmed from nothing but pure weakness, anger, sorrow, and plain stupidity. He asked if he could have a kiss goodbye. She said yes and approached him as he waited by the door. But as she got close, she pretended to gag. She acted with a level of scorn and disgust that was impossible for him to fathom.

Why would I make her want to throw up? Am I that undesirable?

He scolded himself. In that instant, he was unable to recognize any quality in himself. He felt ugly, disgusting, unaccomplished, like a neglected kid.

As he opened the door to allow himself out, from whatever strength he could gather, he said to her as he looked at her for the last time, "I wish I could hurt you as much as you have hurt me. You need professional help. Nobody knows you as I do. Your closest friends, your mom, don't have an idea of the person you really are. To tell you the truth, I can read you like a book, but no one can help you because you're petrified of feeling— Goodbye.

March 27, 2020

As they sit on the grass, Darya listens. The park is a bit more scattered of people at this hour, and the vibrant colors have disappeared from the canvas as the sheet of the darkened sky covered them.

"To be honest with you, I felt like a fool for holding her body and femininity to such regard. At that moment, I understood that her body was more important to me than it had been to her. I couldn't help it nor hold it any longer, and I cried like a baby Darya; I cried, and I cried until I couldn't cry anymore. But there seems to be plenty more" they both chuckle.

"I was so humiliated, degraded, and for what? For patiently waiting for day-long excursions at the mall, in which I ended buying her things she liked? For all the trips, dinners, drinks, outings, and experiences we had? Or was it for being a gentleman? I tried to give her the best boyfriend one could ever have. I tried giving her my love."

"But she cheated on you H., and you shouldn't feel bad. It was a reflection of her, not you."

"Thank you, and I'm not one to justify her, trust me on that. But the truth is I cheated too. As I look back, I cheated myself, and I cheated the person in the mirror. I cheated years of introspection, searching, and growth all because I was afraid of losing something and not getting it back. I managed to deceive myself, and all because I loved who she was on paper, what she validated in my life, despite the constant anxiety I had for months, the passive-aggressive comments like *you're doing well with your little business*' while she couldn't answer a single thing of what she did she liked. *'Speak English then'* because I had mispronounced a word. Or making sly remarks about my own friends *'wow, is there anything that he doesn't do?'* to diminish my accomplishments. Calling me jealous after she made constant innuendos about her friend's boyfriend. *"Trevor and I kissed."* I didn't feel anything by her comment, so I just laughed and joked about it. *Yeah, but you can change in front of your friends babe*, but she **had to** take it further by saying, *"I wouldn't mind him seeing me in underwear. I wonder... no, never mind."* Her last comment obviously bothered me, but it's hard to explain how incredulous it sounded that I didn't know how to react.

But after noticing my discontent, as the weeks went by, she would continue *'you should hear Trevor's position on this matter, I can't explain it. He explains it so advanced,'* or *'I'm surprised you picked a movie with his name,'* when I simply liked the movie because I like ancient Greece and had no idea why it would cause surprise. But overall, most of the time, her comments left me wondering, *why the f*** did I need to know that?*

Making me nearly hate a guy I really liked—a good dude he was. At other times, she would chastise me as being possessive when I wondered where she was after not responding the whole day until the night. Also, often telling me about her sexual experiences in the past, when I hadn't even asked. While denying all functions of intimacy to me. No, I owe an apology to me."

"Sorry H., but what a mind fuck." He can't help but laugh. That's one of the things he loves about this woman, how she manages to challenge him in ways that surprise him, often provoking even a good laugh. "Why do you think she was like that, though?"

"In all sincerity, I just felt she was angry all the time—full of passive aggression, hiding behind a smile, a soft voice, and an act that disappeared as soon as we weren't in the public eye. I genuinely wanted to help *Her* and even thought about telling her friends sometimes, but I knew what that would result in, so I took it all in, hence why I wrote that letter. Talk about self-fulfilling prophecy, huh?"

"Yeah, I remember you telling me about that, when we went to the show at the planetarium when you wouldn't take that picture with me. So what did you end up doing with the letter?"

"I still have it, I stored it away."

"Why? Or for what?"

"I found her mind—*intriguing*—at least different, for sure."

"What do you mean? Different how?

"Different Darya, I don't know, she was different. As different as I have ever seen. She was—vanilla. But never in front of others, never. No one has a clue. The truth is that when we broke up, I wasn't even angry that she did it. I was angry about *how* she did it. I hated the irresponsibility in which she did things, the level of immaturity, callous, malice."

"Didn't you all even hang out that same weekend you came back from El Paso too when I invited you out to the pool?"

"Yeah, with her friends where she boasted about a book I wanted to write. If I would've known, I would have joined you." she subtly interrupts.

"Is it the one on your profile, the one about mythos?"

"Oh wow, you've seen it, thank you. Yes, that one."

"Yeah, I've been through your profile." She chuckles embarrassingly. "So, are you still going to write it?"

"Someday, sure. Now I'm thinking of writing a different one. I need the means to express myself, I have so much I would like to say." She smiles a warm smile that, at that moment, seemed to touch his heart.

That night, once home, a single bedroom apartment felt a bit more spacious than ever before. But it wasn't the night that worried him, since the night had always given him comfort. What he dreaded was the morning, and whatever symptoms awaited. He tried to alleviate the thoughts by lying on his couch while watching a movie. But he saw nothing, only images in his head.

He wasn't sure when the images in his head had turned to dreams, but the dreams were crueler than any vision he inflicted on himself. In his dream, he was still with her, she smiled at him, and he laughed while he teased her. But he can't hear her laughing. So he calls out her name, growing ever more desperate he begins to shout, but he can't hear his words either. Only a song takes over the atmosphere; the music he listened to that day he saw her for the last time. Her face is long, expressionless, and pale as she stares right at him without blinking. He's scared, and his heart begins racing as if knocking to get out. His heart pounds so fast and vicious and jolts him awake.

March 28, 2020

Despite the length of and energy exhausted in the previous days, the clock reads 6:06 a.m. His hands shake, and the anxiety is overwhelming—abruptly dumping acid through his veins and elevating his heartbeat to a sprint. He can see his hands shake as memories attack his mind like a biblical plague that barely allows him to see the room in front of him. His chest is in literal pain, and the sadness is as intense as if a loved one had passed—worse. He reminds himself he's a fighter.

He throws himself to the floor and begins with a set of push-ups. He finds himself unable to stop. The thought of burning the pain away with physical strain comes to mind. He even tries holding his breath as the lactic acid swells his chest and his arms—he nearly passes out. He continues this ritual until he has no more strength to hold back—the tears stream down his eyes. Crying that quickly morphs into sobbing. Sobbing turned into an uncontrollable spasm as he can't seem to stop to cry. As his endless tears dry his eyes, he looks up at the clock, and it reads—6:25 a.m. After thinking he had waged his war for more than an hour, he now gauges a battle minute by minute, and no more than nineteen minutes have passed. His mind is racing, his thoughts are…

The sound of an incoming message buzzes through his headphone piece.

Buenos dias…Omg, I stayed awake, finishing Cora. It was so crazy!!!! Lol, did you finish it? —6:27 a.m.—it's Darya. But H. doesn't waste a heartbeat to respond.

Haha, you're funny. Please tell me you slept lol. I didn't find it, and I really did look. He can't even recall when he turned on the TV, let alone a show he could watch. But the white lie doesn't matter, what matters is keeping Darya's warmth around.

Shit was so intense…

6:22 p.m. later that same day

I'm sorry to say this or bother you... I'm fighting... but I don't feel too good. At this stage, the man has left an orphan child, and the child seeks guidance as he knows not how to fight.

I'm so sorry. You're too nice and smart to be hurting. I'll call you in about an hour, is that cool? And in an hour, she called. What could make her be by his side? He wondered. Having been no more than a bit over twenty-four hours since the onslaught that left him absent of all rational thought, he didn't quite understand Darya's loyalty.

That night his good friend and neighbor Skeet would invite him over for some drinks, a smoke, and a bite. Homemade chicken wings were on the menu, and Skeet had a natural gift in the kitchen, an opportunity H. wasn't about to pass up.

The image H. showed to the world was not the one that he was, but perhaps without knowing it, something had already sparked in him to fight.

March 29th

Hi, how are you today?

Daryaaaa! Como estas hermosa? (How are you doing beautiful?) I've had a much better day today.

I'm so glad. Darya cheers him on.

March 30th

Kelly and I are going to the park today. She's beautiful, and she models, want to come? Darya asks.

Park? Yeah, I'm down. I'm coming around and feeling more and more like H. As if H. was a separate entity from himself, and he refers to him in the third person. Yet, what matters is the pieces he's picking, resembling a person absent or that he thought had gone.

Wow, that was fast. I'm glad. In fact, it *was* fast.

As the days went by, the mornings were bad. He would wake from restless nights to the accelerated rhythm of his heart. Some mornings better than others, but mostly they were tough. But his resilience was tougher and grew fiercer by the hour. And when the symptoms attacked him, he would push back. H. kept himself busy—he built himself a schedule, he got to working, working out, and somehow and for some reason he could not comprehend, day after day Darya stayed by his side.

From moments where they vented for hours to others where they laughed, he had a quarantine to be lonely for hours, but instead, her company is what he got. *What a poetic, surreal, and healing experience,* he thought.

Some days they laughed without an end in sight. Other days, they got drunk while walking, sitting, and talking right there by the park. Sometimes they flustered, but every day without absence in empty streets, they rode their bikes. But regardless of weather, mood, of how tired or sad —they shared many stories, tears, and disagreements alike. But somehow, always writing unknowing a story a page at a time. And as the hours transformed into days, the pandemic moved patiently a day at a time. Yet, whatever experience that happened, it was always—*always*—with each by their side.

April 1st

At 5:00 p.m. at the park? H. asks Darya.

Yep. Where at? Where do we meet? Because I'm about to get my alcohol, lol.

Okay, I'm going too. H. thought she was cuckoo but in the best possible of ways. Because one thing he was sure was that she wasn't joking.

H. can I be 15 minutes late? I'm sorry. I want to walk there since I intend to get fucked up.

Don't worry at all. I'll see you there. I'm arriving. After making it inside and looking for what he could buy, it seemed Darya was always a step ahead, making H. wonder how much more fun she could get:

Cool. I'm drinking already. I'm not a fan of vodka. That day, they would drink at the park. The hours expired, and in laughter, they walked back to the car, as she could no longer hold it, and with a broad smile, they waved at the cops. Once at her place, the recreational hours would last till the night.

April 2ⁿᵈ

H....wtf! OMG! I'm still drunk.

We had fun, don't guilt yourself. Drink some water. H. responds with a headache of his own.

Shit, I did say that. Lol

April 7ᵗʰ

See you in 30—Darya confirms.

Alright, sounds good.

April 15ᵗʰ

Twenty-one days of bike riding with this guy, and we still jump on video calls as we work.—Darya

April 23ʳᵈ

I am indulging in a little treat from my very special someone—H.—Post to his Instagram account.

The days consumed like hours, and slowly but surely, they began living. *And just as time moves fast when you're having fun*, when they least cared to notice, twenty-four days had passed. One day as he woke for his morning hike when the memories were faded and he least expected, as he noticed his symptoms subside, his pain had left him, and *Her* darkness and malice were part of the past.

May 3rd

Hi H., I was wondering, have you heard of the new restaurant they just opened in Galleria? It's Indian cuisine. I've heard from good sources it's the best in the city right now.

Oh yeah? No, I haven't heard of it. But I freaking love Indian food, it's so delicious, and I haven't had it in a while. H. being unaware of the latest and greatest—of anything—was by no stretch surprising. If it wasn't in a book, a scientific journal, or invited by someone, he could barely make out the night from the day; he loved working on his projects and businesses. *Would you like to go?*

Would you take me? She asked. H. bit his lip after reading her message.

Of course, silly goose. I'd love to.

But apparently, it's tough getting a reservation at this place. A friend told me you can't even call.

Oh yeah? How about I call you back? We can both try getting a reservation if you'd like. As he got off the phone, he knew there was no shot in hell he wasn't getting that reservation. *What are friends for?* He thought. Immediately, as a name popped in his head, he wasted no time. He hadn't finished thinking of his full name when he was already calling him.

Darya. We're set for Friday at 7:30 p.m., okay?

Oh, wow, really? How did you do it? He didn't mind her being impressed, as he would admit—*it killed him.* But she would deliver one more devastating blow for good measure.

We should match! We'd look cute together.

I agree. H. responded.

May 7^{th,} 12:41 a.m.

H. entered his home with heavy steps admiring the new image of the man in the mirror as he passed by, and he couldn't help to remember that first day he saw Darya again; it had been a Friday as well. But it had been quite a different one. As he looked back, he wasn't sure if there was a single day he hadn't seen her since. A thought that brought a smile to his face, an expression that looked tired from a night of fun.

Life has an odd sense of humor, he thought. Just a little over a month ago, he found himself shaking in the morning from the anxiety, with a plague of images attacking his mind while he fought to draw them out. Yet here he was, with pictures in his mind from the evening he'd just spend with Darya, fighting to keep them in his mind—*her smell, her laugh, and gosh, the way she looked.* Just thinking about her made him want to put his fist in his mouth, bite hard, and roll his eyes back.

What a year it had been, what a *beautiful* year. He reminisced on the trips, his stepdad, who had just adopted him, and became his dad officially—going through a "delightful" breakup four days before the quarantine, the twenty-sevens. Since then, he had gotten back into his average weight, felt as fresh as a cucumber, learned, worked on his business, and—Darya. Almost as if someone had written it for him. All he could do was smile at the thoughts.

Hmm, an interesting question, he said aloud. He grabbed the first of many dry-erase markers lying around and proceeded with stumbled steps into his office before writing on the board, *what is love?* As expected, with any question that intrigued him, it would soon consume him and produce a thinking frown on his face.

The thoughts and pure emotions he felt at that moment compelled him to tell his story and pour his knowledge on paper. He felt like it had to be at that moment, afraid the experience would fade from his memory, as it naturally does. He had so much to say. Just remembering his grieving gave him goose bumps, and when he tried searching for answers, he hadn't found a single tangible document that guided through the breakup process.

So why not write it myself? He thought people would genuinely benefit from it. Besides being vastly interested in finding out what science had to say about this mysterious phenomenon, people call love. And why did it seem to him like everyone was so afraid of openly expressing it? As he pondered on that last question that came to mind, other questions followed and quickly became five, six, twenty questions lining up to be answered. It excited him.

He knew what a search for answers would be, what a strenuous endeavor he was asking of himself. But now he saw *new* signs. The pandemic was far from gone, he already worked from home, and he happened to have the luxury of having all the time in the world at his disposal. Besides, he had an ocean of information in his brain ready to be downloaded from some time back.

Enthralled by his new idea, he put on a fresh pot of coffee, drank a full glass of water, and headed into his office. He turned on his laptop and pressed the power buttons to the other two monitors. As they came back to live, their glare felt like a greeting saying, "welcome back." "Hi to you too," he said aloud with a wide smile. The coffee finished brewing, and he got up to serve himself a cup. But not without picking a mug from the dozens he owned, appropriate for the occasion—Claude Monet was that night's winner.

Once back at his office chair, he launched Microsoft word, turned on his speaker, and visited his phone to pick music to set the perfect ambiance. He couldn't have chosen anything better to start the night off than Johann Sebastian Bach's Orchestral Suite No 3 in D Major, "Air.[7]"

With his phone still in his hands, he revisited Darya's last message—a kiss emoji. It seemed he couldn't stop smiling; everything inspired him to do so, Darya, the healthy beat of his heart, the ambiance, and his new project—so much happiness roamed the air. *No one should dictate your future, and no one should instill fear in you, poisoning the next great someone. There are people with ill intentions and malice, as there are people with good intentions and kind hearts,* he thought. And the fact that he hadn't allowed himself to fall prey to such illness made him feel proud of himself. That person didn't have a single insult from him, not a single message from him reaching out, not a single act of malice, not a single bad memory, and for that he was content. In that instant, he remembered the words from one of his biggest idols, Martin Luther King, Jr.

"Returning hate for hate multiplies hate, adding deeper darkness to a night already devoid of stars. Darkness cannot drive out darkness; only light can do that. Hate cannot drive out hate; only love can do that."

---

[7] You can visit Hugo Bradford's "Classical Music" playlist on spotify by going to https://open.spotify.com/playlist/49xSUvQmME0RzWCBkxBs8T?si=1rf47eOEQtayzZM M0Mzdjw

As he admired his pictures with Darya from that night, he couldn't help think *I'm happy.* With the cursor blinking at the top of the blank Word page, he typed:

The Anatomy of Love

Bibliography

CHAPTER 1   DEFINING LOVE

PAGE

9   *Interstellar.* Film. USA: Paramount Pictures / Warner Bros., 2014.
262

13   Brogaard, Berit. "Can Animals Love?" Psychology Today. Sussex
      Publishers, February 24, 2014.
      https://www.psychologytoday.com/us/blog/the-mysteries-
      love/201402/can-animals-love.

15   "Pluralistic Ignorance (SOCIAL PSYCHOLOGY) - IResearchNet."
      Psychology, January 21, 2016.
      http://psychology.iresearchnet.com/social-psychology/decision-
      making/pluralistic-ignorance/.

15   College, Reed. "Reed College." Reed College | Pluralistic Ignorance |
      General overview. Accessed August 20, 2020.
      https://www.reed.edu/psychology/pluralisticignorance/index.ht
      ml.

15   Savitsky, Kenneth, and Thomas Gilovich. "The Illusion of
      Transparency and the Alleviation of Speech Anxiety." *Journal of
      Experimental Social Psychology* 39, no. 6 (2003): 618–25.
      https://doi.org/10.1016/s0022-1031(03)00056-8.

22   Reed, Kaye E., John G. Fleagle, and Richard E. Leakey. *The Paleobiology
      of Australopithecus: Contributions from the Fourth Stony Brook Human
      Evolution Symposium and Workshop, Diversity in Australopithecus:
      Tracking the First Bipeds, September 25-28, 2007.* Dordrecht:
      Springer, 2013.

23   Edwards, Scott. "Love and the Brain." Neurobiology. Harvard
      Mahoney Neuroscience Institute, 2020.
      https://neuro.hms.harvard.edu/harvard-mahoney-neuroscience-
      institute/brain-newsletter/and-brain/love-and-brain.

23   Mark, Clifton, and Dr. Mike Dow. "Broken Heart, Broken Brain: The
      Neurology of Breaking up and How to Get over It | CBC Life."
      CBCnews. CBC/Radio Canada, April 6, 2018.

https://www.cbc.ca/life/wellness/broken-heart-broken-brain-the-neurology-of-breaking-up-and-how-to-get-over-it-1.4608785.

CHAPTER 2   BREAKUPS: WHY DO I FEEL THIS WAY?

32   Lachmann Psy.D., Suzanne. "How to Mourn a Breakup to Move Past Grief and Withdrawal." Psychology Today. Sussex Publishers, June 4, 2013. https://www.psychologytoday.com/us/blog/me-we/201306/how-mourn-breakup-move-past-grief-and-withdrawal.

33   Hil, Dr Kim Dr. "Emotions Are Energy : The Bodymind Connection and e-Motion." Authenticity Associates. Dr Kim Dr Hil, January 9, 2020. https://www.authenticityassociates.com/emotions-are-energy/.

34   Orwell, George, Thomas Pynchon, and Erich Fromm. *Nineteen Eighty-Four: a Novel*. New York, NY: Berkley, 2016.

36   Lancer, Darlene. "Is Your Partner Passive-Aggressive?" Psychology
37   Today. Sussex Publishers, June 20, 2017. https://www.psychologytoday.com/us/blog/toxic-relationships/201706/is-your-partner-passive-aggressive.

36   A. Lambert, Carol. "6 Troubling Signs of Psychological Abuse."
37   Psychology Today. Sussex Publishers, August 30, 2017. https://www.psychologytoday.com/us/blog/mind-games/201708/6-troubling-signs-psychological-abuse.

37   Lancer, Darlene. "The Truth About Abusers, Abuse, and What to Do." Psychology Today. Sussex Publishers, June 6, 2017. https://www.psychologytoday.com/us/blog/toxic-relationships/201706/the-truth-about-abusers-abuse-and-what-do.

41   The Editors of Encyclopaedia Britannica. "Serotonin." Encyclopædia
42   Britannica. Encyclopædia Britannica, inc., August 8, 2019. https://www.britannica.com/science/serotonin.

41   "Dopamine." Encyclopædia Britannica. Encyclopædia Britannica, inc., January 18, 2019. https://www.britannica.com/science/dopamine.

41   Rogers, Kara. "Oxytocin." Encyclopædia Britannica. Encyclopædia Britannica, inc., March 20, 2020. https://www.britannica.com/science/oxytocin.

41   Rogers, Robert D. "The Roles of Dopamine and Serotonin in Decision Making: Evidence from Pharmacological Experiments in Humans." *Neuropsychopharmacology* 36, no. 1 (2010): 114–32. https://doi.org/10.1038/npp.2010.165.

41   Svoboda, Elizabeth. "The Thoroughly Modern Guide to Breakups."
      Psychology Today. Sussex Publishers, January 1, 2011.
      https://www.psychologytoday.com/us/articles/201101/the-
      thoroughly-modern-guide-breakups?collection=51836.

42   Schneiderman, Inna, Orna Zagoory-Sharon, James F. Leckman, and
      Ruth Feldman. "Oxytocin during the Initial Stages of Romantic
      Attachment: Relations to Couples' Interactive Reciprocity."
      Psychoneuroendocrinology 37, no. 8 (January 26, 2012): 1–20.
      https://doi.org/10.1016/j.psyneuen.2011.12.021.

42   Gao, Shan, Benjamin Becker, Lizhu Luo, Yayuan Geng, Weihua Zhao,
      Yu Yin, Jiehui Hu, et al. "Oxytocin, the Peptide That Bonds the
      Sexes Also Divides Them." *Proceedings of the National Academy of
      Sciences* 113, no. 27 (2016): 7650–54.
      https://doi.org/10.1073/pnas.1602620113.

43   Greenberg, Melanie. "This Is Your Brain on a Breakup." Psychology
      Today. Sussex Publishers, March 29, 2016.
      https://www.psychologytoday.com/us/blog/the-mindful-self-
      express/201603/is-your-brain-breakup.

43   Kross, E., M. G. Berman, W. Mischel, E. E. Smith, and T. D. Wager.
      "Social Rejection Shares Somatosensory Representations with
      Physical Pain." *Proceedings of the National Academy of Sciences* 108, no.
      15 (2011): 6270–75. https://doi.org/10.1073/pnas.1102693108.

43   Utiger, Robert D. "Vasopressin." Encyclopædia Britannica.
      Encyclopædia Britannica, inc., October 15, 2015.
      https://www.britannica.com/science/vasopressin.

44   Levine, David, and Helen Fisher. "Anthropologist and Love Expert
      Helen Fisher on the Mysteries of Love." Elsevier Connect.
      Elsevier, July 29, 2014.
      https://www.elsevier.com/connect/anthropologist-and-love-
      expert-helen-fisher-on-the-mysteries-of-love.

44   Fisher, Helen & Brown, Lucy & Aron, Arthur & Strong, Greg &
      Mashek, Debra. (2010). Reward, Addiction, and Emotion
      Regulation Systems Associated With Rejection in Love. Journal
      of neurophysiology. 104. 51-60. 10.1152/jn.00784.2009.

45   Fisher, Helen E, Arthur Aron, and Lucy L Brown. "Romantic Love: a
      Mammalian Brain System for Mate Choice." *Philosophical
      Transactions of the Royal Society B: Biological Sciences* 361, no. 1476
      (2006): 2173–86. https://doi.org/10.1098/rstb.2006.1938.

CHAPTER 3   BREAKUPS: COPING 101

47   American Psychological Association. "What Is Cognitive Behavioral
      Theraphy?" American Psychological Association. American

48   Psychological Association, July 2017. https://www.apa.org/ptsd-guideline/patients-and-families/cognitive-behavioral.

48   Gladstein, Randall. *Dane Cook: Vicious Circle. IMDB/ Dane Cook: Vicious Circle.* Home Box Office (HBO) (2006) (USA) (TV), 2006. https://www.imdb.com/title/tt0867149/fullcredits?ref_=ttco_sa_1.

52   Scheve, Tom. "What Are Endorphins?" HowStuffWorks Science. HowStuffWorks, June 30, 2020. https://science.howstuffworks.com/life/inside-the-mind/emotions/endorphins.htm#pt1.

53   Hoyt, Alia. "How Crying Works." HowStuffWorks Science. HowStuffWorks, August 19, 2020. https://science.howstuffworks.com/life/inside-the-mind/emotions/crying.htm.

53   Vingerhoets, Ad J. J. M., and Lauren M. Bylsma. "The Riddle of Human Emotional Crying: A Challenge for Emotion Researchers." *Emotion Review* 8, no. 3 (2016): 207–17. https://doi.org/10.1177/1754073915586226.

53   Gračanin, Asmir, Lauren M. Bylsma, and Ad J. J. M. Vingerhoets. "Why Only Humans Shed Emotional Tears." *Human Nature* 29, no. 2 (2018): 104–33. https://doi.org/10.1007/s12110-018-9312-8.

55   Fernstrom, John D. "Dietary Amino Acids and Brain Function." *Journal of the American Dietetic Association* 94, no. 1 (January 1, 1994): 71–77. https://doi.org/10.1016/0002-8223(94)92045-1.

56   Schäfer, Thomas, Peter Sedlmeier, Christine Städtler, and David Huron. "The Psychological Functions of Music Listening." *Frontiers in Psychology* 4 (August 13, 2013): 1–33. https://doi.org/10.3389/fpsyg.2013.00511.

56   Levitin D. J. (2007). Life Soundtrack: The Uses of Music in Everyday Life. Montreal, QC: McGill University; Available online at: http://levitin.mcgill.ca/pdf/LifeSoundtracks.pdf

56   Bergland, Christopher. "The No. 1 Reason Music Has the Power to Make Us Feel Good." Psychology Today. Sussex Publishers, December 12, 2018. https://www.psychologytoday.com/us/blog/the-athletes-way/201812/the-no-1-reason-music-has-the-power-make-us-feel-good#:~:text=Regarding%20this%20study's%20design%3A%20Participants,sounds%2C%20(3)%20music%20that.

57   Maksimainen, Johanna, Jan Wikgren, Tuomas Eerola, and Suvi Saarikallio. "The Effect of Memory in Inducing Pleasant

Emotions with Musical and Pictorial Stimuli." *Scientific Reports* 8, no. 1 (November 7, 2018): 1–12. https://doi.org/10.1038/s41598-018-35899-y.

58   Berkheiser, Kaitlyn. "The 6 Best Bedtime Teas That Help You Sleep." Healthline. Healthline Media, October 21, 2019. https://www.healthline.com/nutrition/teas-that-help-you-sleep#1.

58   Gupta, Sanjay, Eswar Shankar, and Janmejai K Srivastava. "Chamomile: A Herbal Medicine of the Past with a Bright Future (Review)." *Molecular Medicine Reports* 3, no. 6 (November 28, 2010). https://doi.org/10.3892/mmr.2010.377.

59   Mineo, Liz. "Less Stress, Clearer Thoughts with Mindfulness
60   Meditation." Harvard Gazette. Harvard Gazette, November 1, 2019. https://news.harvard.edu/gazette/story/2018/04/less-stress-clearer-thoughts-with-mindfulness-meditation/.

62   Mark, Clifton, and Dr. Mike Dow. "Broken Heart, Broken Brain: The Neurology of Breaking up and How to Get over It | CBC Life." CBCnews. CBC/Radio Canada, April 6, 2018. https://www.cbc.ca/life/wellness/broken-heart-broken-brain-the-neurology-of-breaking-up-and-how-to-get-over-it-1.4608785.

64   University of Colorado at Boulder. "Your brain on imagination: It's a lot like reality, study shows." ScienceDaily. www.sciencedaily.com/releases/2018/12/181210144943.htm (accessed August 20, 2020).

65   Reddan, Marianne Cumella, Tor Dessart Wager, and Daniela Schiller. "Attenuating Neural Threat Expression with Imagination." *Neuron* 100, no. 4 (November 21, 2018): 1–36. https://doi.org/10.1016/j.neuron.2018.10.047.

CHAPTER 4   WHAT YOU SAY YOU WANT

73   *Bedazzled. IMDB / Bedazzled.* Twentieth Century Fox, 2000.
292   https://www.imdb.com/title/tt0230030/?ref_=ttco_co_tt.

75   Seuss, Dr. *Oh, the Places You Will Go.* Random House, NY: New York, 1990.

75   Max Roser, Esteban Ortiz-Ospina and Hannah Ritchie (2013) - "Life Expectancy". Published online at OurWorldInData.org. Retrieved from: 'https://ourworldindata.org/life-expectancy' [Online Resource]

76   Degges-White, Suzanne. "15 Things Women Want From the Men in
77   Their Lives." Psychology Today. Sussex Publishers, June 4, 2018. https://www.psychologytoday.com/us/blog/lifetime-

connections/201806/15-things-women-want-the-men-in-their-lives.

76
77
Jones, Alexis. "10 Of the Most Important Qualities Women Look for in a Guy." Redbook. Redbook, October 23, 2019. https://www.redbookmag.com/love-sex/relationships/a22750311/what-qualities-women-want-in-a-man/.

77　Chernin, Peter, Dylan Clark, Rick Jaffa, Rick Jaffa, Amanda Silver, Amanda Silver, and Mark Bomback. *Dawn of the Planet of the Apes*. United States: 20th Century Fox, 2014.

78　Pychyl, Timothy A. "Hierarchy of Excuses: The Pathetic Path of Least Resistance." Psychology Today. Sussex Publishers, March 18, 2011. https://www.psychologytoday.com/us/blog/dont-delay/201103/hierarchy-excuses-the-pathetic-path-least-resistance.

78　Cherry, Kendra, and Steven Gans, MD. "Cognitive Dissonance and Ways to Resolve It." Verywell Mind. Verywell Mind, July 2, 2020. https://www.verywellmind.com/what-is-cognitive-dissonance-2795012.

79　Denholm Ph.D.., L.M.H.C., Diana B. "Man the Fixer, Woman the Nurturer-the Caregiving Gender Gap." Psychology Today. Sussex Publishers, April 23, 2012. https://www.psychologytoday.com/us/blog/the-caregivers-handbook/201204/man-the-fixer-woman-the-nurturer-the-caregiving-gender-gap.

80　Chalabi, Mona. "What's The Average Age Difference In A Couple?" FiveThirtyEight. FiveThirtyEight, January 22, 2015. https://fivethirtyeight.com/features/whats-the-average-age-difference-in-a-couple/.

85　Geographic, National. "Black Widow Spiders." National Geographic. National Geographic, September 24, 2018. https://www.nationalgeographic.com/animals/invertebrates/group/black-widow-spiders/.

85　Fullingim, Photograph by Sharon. "It's Praying Mantis Mating Season: Here's What You Need To Know." What to Know for Praying Mantis Mating Season. National Geographic, September 7, 2018. https://www.nationalgeographic.com/animals/2018/09/praying-mantis-mating-cannibalism-birds-bite-facts-news/.

85　Fullingim, Photograph by Sharon. "It's Praying Mantis Mating Season: Here's What You Need To Know." What to Know for Praying Mantis Mating Season. National Geographic, September 7, 2018.

https://www.nationalgeographic.com/animals/2018/09/praying
-mantis-mating-cannibalism-birds-bite-facts-news/.

90   Darwin, Charles, Paul H. Barrett, and R. B. Freeman. *The Descent of Man, and Selection in Relation to Sex.* London, UK: Routledge, 2017.

90   Wilson, Michael Lawrence, Carrie M. Miller, and Kristin N. Crouse. "Humans as a Model Species for Sexual Selection Research." *Proceedings of the Royal Society B: Biological Sciences* 284, no. 1866 (October 10, 2017): 1–10. https://doi.org/10.1098/rspb.2017.1320.

91   Jewell, Tim. "How Long Does It Take for Sperm to Regenerate? Tips for Production." Healthline. Healthline Media, September 19, 2018. https://www.healthline.com/health/mens-health/how-long-does-it-take-for-sperm-to-regenerate#sperm-production-rate.

91   Bullivant, Susan B., Sarah A. Sellergren, Kathleen Stern, Natasha A. Spencer, Suma Jacob, Julie A. Mennella, and Martha K. Mcclintock. "Women's Sexual Experience during the Menstrual Cycle: Identification of the Sexual Phase by Noninvasive Measurement of Luteinizing Hormone." *Journal of Sex Research* 41, no. 1 (February 2004): 82–93. https://doi.org/10.1080/00224490409552216.

93   Mosher WD, Chandra A, Jones J. Sexual behavior and selected health measures: Men and women 15–44 years of age, United States, 2002 . Advance data from vital and health statistics; no 362. Hyattsville, MD: National Center for Health Statistics. 2005.

93   Smith, Ph.D.., Drew. "How Many Possible Combinations Of DNA Are There?" Forbes. Forbes Magazine, January 20, 2017. https://www.forbes.com/sites/quora/2017/01/20/how-many-possible-combinations-of-dna-are-there/#2895c7285835.

CHAPTER 5   WHAT WOMEN REALLY WANT

98   Harari, Yuval Noah. *Sapiens: a Brief History of Humankind.* New York, NY: Harper Perennial, 2018.

100   Cherry, Kendra, and David Susman Ph.D.. "How Does the Weschsler
101   Adult Intelligence Scale Measure Intelligence?" Verywell Mind. 2020 About, Inc. , February 19, 2020. https://www.verywellmind.com/the-wechsler-adult-intelligence-scale-2795283.

100   Jones, Alexis. "10 Of the Most Important Qualities Women Look for
115   in a Guy." Redbook. Redbook, October 23, 2019. https://www.redbookmag.com/love-

116   sex/relationships/a22750311/what-qualities-women-want-in-a-
         man/.

101   Denney, David A., Wendy K. Ringe, and Laura H. Lacritz. "Dyadic
         Short Forms of the Wechsler Adult Intelligence Scale-IV."
         Archives of Clinical Neuropsychology 30, no. 5 (2015): 404–12.
         https://doi.org/10.1093/arclin/acv035.

104   Eagly, Alice H., and Wendy Wood. "The Origins of Sex Differences in
         Human Behavior: Evolved Dispositions versus Social Roles."
         American Psychologist 54, no. 6 (1999): 408–23.
         https://doi.org/10.1037/0003-066x.54.6.408.

104   Zhang, Lingshan, Anthony J. Lee, Lisa M. Debruine, and Benedict C.
         Jones. "Are Sex Differences in Preferences for Physical
         Attractiveness and Good Earning Capacity in Potential Mates
         Smaller in Countries With Greater Gender Equality?" *Evolutionary
         Psychology* 17, no. 2 (2019): 147470491985292.
         https://doi.org/10.1177/1474704919852921.

104   Cafasso, Jacquelyn. "Average IQ: US, Globally, How It's Measured,
         and Controversies." Healthline. Healthline Media, April 10, 2018.
         https://www.healthline.com/health/average-iq#takeaway.

105   Gensowski, Miriam. Rep. *Personality, IQ, and Lifetime Earnings*. Bonn,
         Germany: IZA, 2014.

107   Pierre, Joseph M. "Why Do We Cry? Exploring the Psychology of
         Emotional Tears." Psychology Today. Sussex Publishers, April
         23, 2018. https://www.psychologytoday.com/us/blog/psych-
         unseen/201804/why-do-we-cry-exploring-the-psychology-
         emotional-tears.

111   Rampton, Martha. "Four Waves of Feminism." Pacific University.
         Pacific magazine, July 13, 2020.
         https://www.pacificu.edu/magazine/four-waves-feminism.

116   Fisher, Dr Helen. "We have chemistry ! – the role of four primary
         temperament dimensions in mate choice and partner
         compatibility." (2012).

CHAPTER 6   WHAT MAKES US COMPATIBLE?

121   Ackerman, MSc., Courtney E. "Big Five Personality Traits: The
         OCEAN Model Explained [2019 Upd.]."
  -      PositivePsychology.com. Positive Psychology, April 10, 2020.
123   https://positivepsychology.com/big-five-personality-theory/.

126

124   John, O. P., & Srivastava, S. (1999). The Big-Five trait taxonomy:
         History, measurement, and theoretical perspectives. In L. A.

125   Pervin & O. P. John (Eds.), Handbook of personality: Theory
           and research (Vol. 2, pp. 102–138). New York: Guilford Press

125   Power, R A, and M Pluess. "Heritability Estimates of the Big Five
       Personality Traits Based on Common Genetic Variants."
       Translational Psychiatry 5, no. 7 (July 14, 2015): 1–4.
       https://doi.org/10.1038/tp.2015.96.

125   Jang, Kerry L., W. John Livesley, and Philip A. Vemon. "Heritability of
       the Big Five Personality Dimensions and Their Facets: A Twin
       Study." *Journal of Personality* 64, no. 3 (September 1996): 577–92.
       https://doi.org/10.1111/j.1467-6494.1996.tb00522.x.

125   Chapman, Benjamin P., Paul R. Duberstein, Silvia Sörensen, and
130   Jeffrey M. Lyness. "Gender Differences in Five Factor Model
           Personality Traits in an Elderly Cohort." *Personality and Individual*
131   *Differences* 43, no. 6 (October 2007): 1594–1603.
           https://doi.org/10.1016/j.paid.2007.04.028.

125   Weisberg, Yanna J., Colin G. Deyoung, and Jacob B. Hirsh. "Gender
130   Differences in Personality across the Ten Aspects of the Big
           Five." *Frontiers in Psychology* 2 (August 2011): 1–11.
131   https://doi.org/10.3389/fpsyg.2011.00178.

126   Cherry, Kendra, and Steven Gans, MD. "What Are the Big 5
 –   Personality Traits?" Verywell Mind. Verywell Mind, July 13, 2020.
132   https://www.verywellmind.com/the-big-five-personality-
           dimensions-2795422.

126   McGreal, Scott. "What Is An Intelligent Personality?" Psychology
127   Today. Sussex Publishers, November 3, 2014.
           https://www.psychologytoday.com/us/blog/unique-everybody-
132   else/201411/what-is-intelligent-personality.

127   *The Matrix*. Film. USA: Warner Bros., 1999.

218

131   Costa, Paul T., Antonio Terracciano, and Robert R. Mccrae. "Gender
       Differences in Personality Traits across Cultures: Robust and
       Surprising Findings." *Journal of Personality and Social Psychology* 81,
       no. 2 (August 2001): 322–31. https://doi.org/10.1037/0022-
       3514.81.2.322.

131   Judge, Timothy A., and Remus Ilies. "Relationship of Personality to
       Performance Motivation: A Meta-Analytic Review." *Journal of*
       Applied Psychology 87, no. 4 (August 2002): 797–807.
       https://doi.org/10.1037/0021-9010.87.4.797.

132   Pychyl, Timothy A. "Hierarchy of Excuses: The Pathetic Path of Least
       Resistance." Psychology Today. Sussex Publishers, March 18,
       2011. https://www.psychologytoday.com/us/blog/dont-

delay/201103/hierarchy-excuses-the-pathetic-path-least-resistance.

132 Cherry, Kendra, and Steven Gans, MD. "Cognitive Dissonance and Ways to Resolve It." Verywell Mind. Verywell Mind, July 2, 2020. https://www.verywellmind.com/what-is-cognitive-dissonance-2795012.

133 Schretlen, David J., Egberdina-Józefa Van Der Hulst, Godfrey D. Pearlson, and Barry Gordon. "A Neuropsychological Study of Personality: Trait Openness in Relation to Intelligence, Fluency, and Executive Functioning." *Journal of Clinical and Experimental Neuropsychology* 32, no. 10 (April 19, 2010): 1068–73. https://doi.org/10.1080/13803391003689770.

133 Malouff, John M., Einar B. Thorsteinsson, and Nicola S. Schutte. "The Relationship Between the Five-Factor Model of Personality and Symptoms of Clinical Disorders: A Meta-Analysis." *Journal of Psychopathology and Behavioral Assessment* 27, no. 2 (June 2005): 101–14. https://doi.org/10.1007/s10862-005-5384-y.

CHAPTER 7 WHAT YOU DESERVE

143   Festinger, Leon. *A Theory of Cognitive Dissonance*. Stanford, CA: Stanford
144   University Press, 2009.

143   Vaidis, David C., and Alexandre Bran. "Respectable Challenges to
144   Respectable Theory: Cognitive Dissonance Theory Requires Conceptualization Clarification and Operational Tools." *Frontiers in Psychology* 10 (2019). https://doi.org/10.3389/fpsyg.2019.01189.

143   Cancino-Montecinos, Sebastian, Fredrik Björklund, and Torun
144   Lindholm. "Dissonance Reduction as Emotion Regulation: Attitude Change Is Related to Positive Emotions in the Induced Compliance Paradigm." *Plos One* 13, no. 12 (2018). https://doi.org/10.1371/journal.pone.0209012.

145   Pychyl, Timothy A. "Hierarchy of Excuses: The Pathetic Path of Least
146   Resistance." Psychology Today. Sussex Publishers, March 18, 2011. https://www.psychologytoday.com/us/blog/dont-delay/201103/hierarchy-excuses-the-pathetic-path-least-resistance.

147   Report, Emory. "What the Declaration of Independence Really Means by 'Pursuit of Happiness'." Emory University | Atlanta, GA. Emory News Center, July 3, 2018. https://news.emory.edu/stories/2014/06/er_pursuit_of_happiness/campus.html.

154   Ackerman, MSc, Courtney E. "What Is Self-Worth and How Do We
       Increase It? (Incl. 4 Worksheets)." PositivePsychology.com.
       Positive Psychology, April 15, 2020.
       https://positivepsychology.com/self-worth/.

CHAPTER 8   WHAT MEN THINK

161   Galilei, Galileo. Letter to Christina of Tuscany. "Letter to the Grand
       Duchess Christina of Tuscany ." Florence, Tuscany: Italy, 1615.

166   Cherry, Kendra, and Amy Morin, LCSW. "The Basics of Prosocial
–   Behavior." Verywell Mind. Verywell Mind, April 30, 2020.
168   https://www.verywellmind.com/what-is-prosocial-behavior-
       2795479#:~:text=In%20The%20Handbook%20of%20Social,co
       mforting%2C%20sharing%20and%20cooperation.%22.

167   Ueno, Hiroshi, Shunsuke Suemitsu, Shinji Murakami, Naoya Kitamura,
       Kenta Wani, Yu Takahashi, Yosuke Matsumoto, Motoi
       Okamoto, and Takeshi Ishihara. "Rescue-like Behaviour in Mice
       Is Mediated by Their Interest in the Restraint Tool." *Scientific
       Reports* 9, no. 1 (2019). https://doi.org/10.1038/s41598-019-
       46128-5.

167   Mitchum, Rob. "Helping Your Fellow Rat: Rodents Show Empathy-
       Driven Behavior." University of Chicago News. uchicaco news,
       December 8, 2011. https://news.uchicago.edu/story/helping-
       your-fellow-rat-rodents-show-empathy-driven-behavior.

169   Naumann, Carl, Elizabeth Naumann, and LLoyd Minor, MD. Letter to
175   Board. "Nature, Nurture, Sex and Gender." Redwood City, CA:
       Stanford University School of Medicine, n.d.

172   Mcrae, Kateri, Kevin N. Ochsner, Iris B. Mauss, John J. D. Gabrieli,
       and James J. Gross. "Gender Differences in Emotion Regulation:
       An FMRI Study of Cognitive Reappraisal." *Group Processes &
       Intergroup Relations* 11, no. 2 (2008): 143–62.
       https://doi.org/10.1177/1368430207088035.

CHAPTER 9   WHAT MENT WANT

176   McLeod, S. A. (2019, July 18). Psychosexual stages. Simply Psychology.
178   https://www.simplypsychology.org/psychosexual.html

177   Encyclopaedia Britannica, The Editors of. "Oedipus." Encyclopædia
       Britannica. Encyclopædia Britannica, inc., February 5, 2020.
       https://www.britannica.com/topic/Oedipus-Greek-mythology.

178   Digamon, Jayson S., Reymarvelos M. Oros, Zerg D. Encenzo, Blessed
       Joana O.takling, Rynia Mariastelle I.apugan, John Dave
       Lingatong, and Jericho O. Pon. "The Role of the Oedipus

Complex on the Perceived Romantic Security of Males." *International Journal of Scientific and Research Publications (IJSRP)* 9, no. 4 (April 12, 2019): 1–12. https://doi.org/10.29322/ijsrp.9.04.2019.p8858.

181   Jung, C. G. *Two Essays on Analytical Psychology*. Princeton, , NJ: Princeton University Press, 1966.

181   Fischer-Shofty, Meytal, Yechiel Levkovitz, and Simone G. Shamay-Tsoory. "Oxytocin Facilitates Accurate Perception of Competition in Men and Kinship in Women." *Social Cognitive and Affective Neuroscience* 8, no. 3 (2012): 313–17. https://doi.org/10.1093/scan/nsr100.

181   Naumann, Carl, Elizabeth Naumann, and LLoyd Minor, MD. Letter to Board. "Nature, Nurture, Sex and Gender." Redwood City, CA: Stanford University School of Medicine, n.d.

184   Jaffe, Eric. "What Do Men Really Want?" Psychology Today. Sussex Publishers, March 13, 2012. https://www.psychologytoday.com/us/articles/201203/what-do-men-really-want.

184 – 190   Otoya, Lana. "What Men Look For In Women: From A Professional Matchmaker -." Millennialships Dating, August 6, 2020. https://millennialships.com/what-men-look-for-in-women/.

186   Cherry, Kendra. "What Are Fluid Intelligence and Crystallized Intelligence?" Verywell Mind. Verywell Mind, December 7, 2019. https://www.verywellmind.com/fluid-intelligence-vs-crystallized-intelligence-2795004#:~:text=Crystallized%20intelligence%20refers%20to%20the,to%20decline%20as%20they%20age.

188 – 189   Reynolds, Marcia. "What Does It Mean to Be Feminine?" Psychology Today. Sussex Publishers, December 13, 2010. https://www.psychologytoday.com/us/blog/wander-woman/201012/what-does-it-mean-be-feminine.

CHAPTER 10 DATING

194   Smith, Eric Alden, Kim Hill, Frank W. Marlowe, David Nolin, Polly Wiessner, Michael Gurven, Samuel Bowles, Monique Borgerhoff Mulder, Tom Hertz, and Adrian Bell. "Wealth Transmission and Inequality among Hunter-Gatherers." *Current Anthropology* 51, no. 1 (February 2010): 19–34. https://doi.org/10.1086/648530.

198   *Days of Thunder*. Film. USA: Paramount Pictures / Don Simpson/Jerry Bruckheimer Films, 1990.

204   Noggle, Robert, "The Ethics of Manipulation", The Stanford
       Encyclopedia of Philosophy (Summer 2020 Edition), Edward N.
       Zalta (ed.), URL = .

CHAPTER 11 THE HONEYMOON STAGE

213   *The Wizard of Oz*. Film. United States: Metro-Goldwyn-Mayer (MGM),
       1990.

216   Mark, Clifton, and Dr. Mike Dow. "Broken Heart, Broken Brain: The
       Neurology of Breaking up and How to Get over It | CBC Life."
       CBCnews. CBC/Radio Canada, April 6, 2018.
       https://www.cbc.ca/life/wellness/broken-heart-broken-brain-
       the-neurology-of-breaking-up-and-how-to-get-over-it-1.4608785.

217   Petruzzello, Melissa. "Understanding Newton's Laws of Motion."
       Encyclopædia Britannica. Encyclopædia Britannica, inc., 2020.
       https://www.britannica.com/story/understanding-newtons-laws-
       of-
       motion#:~:text=The%20Third%20Law,equal%20and%20opposi
       te%20reaction%E2%80%9D).

CHAPTER 12 THE FOUNDATION: COMMUNICATION

221   Chapman, Gary D. *The 5 Love Languages*. Chicago, IL: Northfield Pub.,
       2015.

222   *Bridesmaids*. Film. USA: Universal, 2011.

225   Conte, Christian. Dr. Christian Conte. Dr. Christian Conte, 2020.
       http://www.drchristianconte.com/about/.

234   Ohlin, MA, BBA, Birgit. "7 Ways to Improve Communication in
       Relationships [Update 2019]." PositivePsychology.com. Positive
       Psychology, April 17, 2020.
       https://positivepsychology.com/communication-in-
       relationships/.

CHAPTER 13 THE FOUR ELEMENTS

239   Home Science Tools. "The Four Elements of Matter: Earth, Water,
       Air, Fire." Home Science Tools. Home Science Tools, September
       24, 2019. https://learning-
       center.homesciencetools.com/article/four-elements-
       science/#:~:text=The%20ancient%20Greeks%20believed%20th
       at,and%20added%20to%20by%20Aristotle.

241   Betchen, Stephen J. "The Importance of Mutual Respect in Intimate
–   Relationships." Psychology Today. Sussex Publishers, March 25,

245   2019. https://www.psychologytoday.com/us/blog/magnetic-partners/201903/the-importance-mutual-respect-in-intimate-relationships.

241   Gray, Peter. "In Relationships, Respect May Be Even More Crucial
–   Than Love." Psychology Today. Sussex Publishers, August 19,
245   2012. https://www.psychologytoday.com/us/blog/freedom-learn/201208/in-relationships-respect-may-be-even-more-crucial-love.

241   Bloom, Linda. "25 Ways You Can Show Respect to Your Partner."
–   Psychology Today. Sussex Publishers, January 25, 2017.
245   https://www.psychologytoday.com/us/blog/stronger-the-broken-places/201701/25-ways-you-can-show-respect-your-partner.

241   Huo, Yuen J., and Kevin R. Binning. "Why the Psychological
–   Experience of Respect Matters in Group Life: An Integrative
245   Account." *Social and Personality Psychology Compass* 2, no. 4 (2008): 1570–85. https://doi.org/10.1111/j.1751-9004.2008.00129.x.

246   Zhang, Shuangyue. "Is Honesty the Best Policy? Honest but Hurtful
247   Evaluative Messages in Romantic Relationships."
            Https://Etd.ohiolink.edu/ IS HONESTY THE BEST POLICY? HONEST BUT HURTFUL EVALUATIVE MESSAGES IN ROMANTIC RELATIONSHIPS. Dissertation, The Ohio State University, 2005.
            https://etd.ohiolink.edu/!etd.send_file?accession=osu1123853679&disposition=inline.

246   Goldsmith, Barton. "Honesty Can Make or Break a Relationship."
247   Psychology Today. Sussex Publishers, November 5, 2014.
            https://www.psychologytoday.com/us/blog/emotional-fitness/201411/honesty-can-make-or-break-relationship.

246   7th, bella September, and Miss lis March 27th. "Do You Have an
247   Honest Relationship?" PsychAlive. PsychAlive, June 8, 2015.
            https://www.psychalive.org/do-you-have-an-honest-relationship/.

246   Debnam, Katrina J., Donna E. Howard, and Mary A. Garza. "'If You
247   Don't Have Honesty in a Relationship, Then There Is No
            Relationship': African American Girls' Characterization of Healthy Dating Relationships, A Qualitative Study." *The Journal of Primary Prevention* 35, no. 6 (2014): 397–407.
            https://doi.org/10.1007/s10935-014-0362-3.

248   Brogaard, Berit. "5 Signs That You're Dealing With a Passive-
–   Aggressive Person." Psychology Today. Sussex Publishers,
250   November 13, 2016.

https://www.psychologytoday.com/us/blog/the-superhuman-mind/201611/5-signs-youre-dealing-passive-aggressive-person.

251 Goldsmith, Barton. "10 Reasons to C.O.M.P.R.O.M.I.S.E. in Your
– Relationship." Psychology Today. Sussex Publishers, April 19,
254 2018. https://www.psychologytoday.com/us/blog/emotional-fitness/201804/10-reasons-compromise-in-your-relationship.

251 Seltzer, Leon F. "Compromise Made Simple: 7 Handy Tips for
– Couples." Psychology Today. Sussex Publishers, October 29,
254 2015. https://www.psychologytoday.com/us/blog/evolution-the-self/201510/compromise-made-simple-7-handy-tips-couples.

251 McKay, Matthew, Peter D. Rogers, and Judith McKay. *When Anger
– Hurts Quieting the Storm Within*. Oakland, CA: New Harbinger,
254 2003.

254 Craig, BPsySc, Heather. "10 Ways To Build Trust in a Relationship."
PositivePsychology.com. Positive Psychology, April 10, 2020.
https://positivepsychology.com/build-trust/.

255 *Ransom*. Film. USA: Touchstone Pictures / Buena Vista Pictures, 1956.

CHAPTER 14 THE FIFTH ELEMENT

260 Safron, Adam. "What Is Orgasm? A Model of Sexual Trance and
270 Climax via Rhythmic Entrainment." *Socioaffective Neuroscience &
Psychology* 6, no. 1 (August 5, 2016): 1–17.
https://doi.org/10.3402/snp.v6.31763.

260 Moore, Psychologist, MBA, Catherine. "What Is Flow in Psychology?
Definition and 10+ Activities to Induce Flow."
PositivePsychology.com. Positive Psychology, May 20, 2020.
https://positivepsychology.com/what-is-flow/.

260 Csikszentmihalyi, Mihaly. *Finding Flow: the Psychology of Engagement with
Everyday Life*. New York, NY, NY: Basic Books, 2008.

261 Naumann, Carl, Elizabeth Naumann, and LLoyd Minor, MD. Letter to
Board. "Nature, Nurture, Sex and Gender." Redwood City, CA:
Stanford University School of Medicine, n.d.

262 Safron, Adam. "What Is Orgasm? A Model of Sexual Trance and
Climax via Rhythmic Entrainment." *Socioaffective Neuroscience &
Psychology* 6, no. 1 (March 29, 2016): 1–18.
https://doi.org/10.3402/snp.v6.31763.

269 Ayala, Francisco Jose. "Sexual Selection." Encyclopædia Britannica.
Encyclopædia Britannica, inc., August 8, 2019.
https://www.britannica.com/science/sexual-selection.

The Anatomy of Love 344

271   Alcock, J., KM. Noonan R. Alexander, WB. Lemmon ML. Allen, M.
273   Andersson, M. Apostolou, MA. Bellis RR. Baker, D. Barash, et al.
           "Why Women Have Orgasms: An Evolutionary Analysis."
           Archives of Sexual Behavior. Springer US, January 1, 1980.
           https://link.springer.com/10.1007/s10508-012-9967-x.

273   Sherlock, James M., Morgan J. Sidari, Emily Ann Harris, Fiona Kate
           Barlow, and Brendan P. Zietsch. "Testing the Mate-Choice
           Hypothesis of the Female Orgasm: Disentangling Traits and
           Behaviours." *Socioaffective Neuroscience & Psychology* 6, no. 1
           (October 25, 2016): 1–10.
           https://doi.org/10.3402/snp.v6.31562.

275   Wakin, A., & Vo, D. B. (2008). Love-variant: The Wakin-Vo I. D. R.
           model of limerence. Inter-Disciplinary – Net. 2nd
           GlobalConference;Challenging Intimate Boundaries. Retrieved 13
           February 2019

275   Keller M.A., Kristine. "Limerence: When Is It More than Heartbreak?"
           Psychology Today. Sussex Publishers, September 23, 2011.
           https://www.psychologytoday.com/us/blog/the-young-and-the-
           restless/201109/limerence-when-is-it-more-heartbreak.

275   Mitrokostas, Sophia. "Here's What Happens to Your Body And Brain
           When You Orgasm." ScienceAlert. Business Insider, January 26,
           2019. https://www.sciencealert.com/here-s-what-happens-to-
           your-brain-when-you-orgasm.

278   Goldstein MD, Irwin. "Orgasmic Anhedonia (PDOD)." Orgasmic
           Anhedonia (PDOD) | San Diego Sexual Medicine. San Diego
           Sexual Medicine, 2020.
           http://sandiegosexualmedicine.com/female-issues/orgasmic-
           anhedonia-pdod.

281   Durairajanayagam, Damayanthi & Rengan, Anil & Sharma, Rakesh &
           Agarwal, Ashok. (2015). Sperm Biology from Production to
           Ejaculation. 10.1007/978-1-4939-2140-9_5.

CHAPTER 15 A LOVE STORY OR A STORY ABOUT LOVE

307   Namka Ed. D., Lynne. "Abandonment Rage." Lynne Namka.
           CONSCIOUS CLARITY CENTER INC., October 20, 2016.
           https://lynnenamka.com/abandonment/abandonment-
           articles/abandonment-rage/.

Notes

July 27, 2020

"I think I'll have a vegan burrito. Want to come with?" H. tells the lovely lady on the phone—it's his mother. She laughs at his question, which has become an inside joke at this point. She lives 750 miles away, and whenever he takes a break from work, which is usually when he heads to grab a bite, he often calls her to say hi, asking if she wants to accompany him.

"Sure thing, where are we going this time, son? Are you going to try eating vegan again?" She refers to a period of two weeks that he gave the vegan life a try.

"No, not at all. It's interesting, perhaps in the future, but you know how busy I am. No, Freebirds just happens to have one mean vegan burrito. Besides, I haven't had one in over four months." H. was introduced to this deli burrito by his ex, but it'd been a while since he had it, since it wasn't the healthiest of meals. Yet, the past was the past, he thought, and he'd shed off some extra pounds in the last couple of months, so he convinced himself with a nice and stern "meh, why not?" The way he saw it, he was, but a few days from his birthday, he also received a birthday present from his parents that same day, and another pretty lady in his life had given him a gift as well.

"My editor said she'd give me the book back on the 31st."

"oh wow, really, on your birthday? That's great, what a birthday gift."

"I know, right? I'm super excited. I'm also super excited about this burrito I'm about to inhale because I haven't had anything all day, and after the 31st, I'm going to be completely plugged into my work."

"I know you will, so I'm happy you're treating yourself. Isn't today when you celebrated your birthday last year?" Once he parked and was halfways towards the entrance, he came to a sudden halt, remembering he'd forgotten his face mask—as he always did.

"It is. Ugh, this damn thing, I always forget it, it's in the car, though, thankfully." He quickly snatches it from the glove compartment and heads back over to the building anticipating his burrito. "Imagine if... wait, it looks like... get the fuck outta here, it's her."

"Who? What happened?" As he tries to give her an answer, he quickly runs back to his car and puts on the front windshield sun covers. It was his ex, who he hadn't seen since their breakup, heading out of the restaurant. Waiting in his car, he saw her from the corner of his eye, analyzing his vehicle while he stared at his phone. He waited until she made it to her car before getting out of his.

"oh my gosh, did she see you?"

"She saw the car, that's for sure."

"oh wow, what are the chances of that?" they both burst out a laugh. As their laughing ceases, with a smile remaining on his face, he says to himself: *huh, life has an odd sense of humor.*

Trip to Phoenix   6.65mph = 9:01mp/mile

Scottsdale
· Taliesin West
· Old Town Scottsdale
— Shade Rooftop Bar

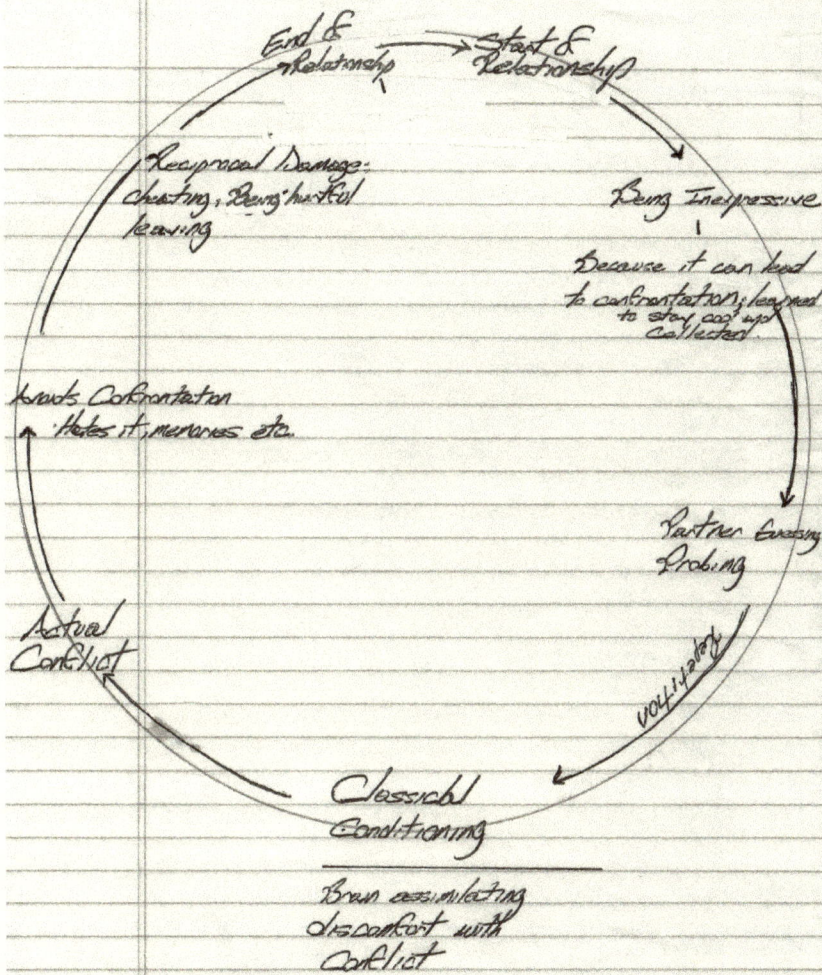

End of
Relationship → Start of
Relationship

Reciprocal Damage:
cheating, Being hurtful
leaving

Being Inexpressive

Because it can lead
to confrontation; learned
to stay cool and
collected

Brain's Confrontation
- Hates it, memories etc.

Partner Growing
Probing

Actual
Conflict

voulytation

Classical
Conditioning

_____

Brain associating
discomfort with
Conflict

Thursday 16th
We arrive @ 10:40 PM
- we ~~route~~ go to bed to go hiking
early.

Friday 17th
- Hiking @ 6AM
- @ 10 AM option of Breakfast Burritos

<u>Cons Meets Awareness</u>

Design
- Human ~~Biology~~ Male
Equation-Golden ratio - Higher levels of
- Penis Shape Competition. tests.
- Human Pheromone - Higher levels of
- Menstrual Cycle estrogen
- Sexual Selection - Strength, Shoulders etc.

learning
It is only after ~~acquiring~~ this knowledge
and acquiring this level of understanding
that we are free to dissect our original
list of preferences.
~~Effectively/Efficiently/Correctly/consciously~~
Survival of the fittest would serve to
exist.

...know is that upon the introduction of
~~consciousness~~ the cycle illustrated/explained
above forever changed. ~~It ceases~~

- How So?
  + Eradicated species - unprecedented level of
    violence/ Life on the planet hadn't evolved to be
    prepared for the introduction of humans.
  + Jumping to the top of the food chain.
  +
  + # For the first time a living species would
    go from surviving to complete domination/Apathetic
    and for the first time (new record) the
    weak were not destined to die but be dominated
    by the strong - ever so confused about
    their purpose in life, depressed, lost and confused not
    our nature and ... unloved. Ever conforming to
    selection ... the currents without realizing the word
    between being affected or unfazed by those currents
    is a matter of choice / because regardless of each ...
  → mindful - don't try - conscious choosing not to be aware.
    - Approach the Q "what is love?" from
    a metaphysical perspective

Methodology
4 Action words : Understanding, Learning,
Accepting, Applying.

4 Tools to implement and motivate those
actions : Honesty, Perspective, Courage,
willingness.

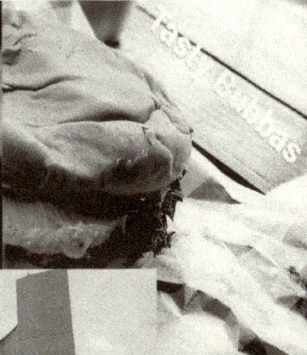

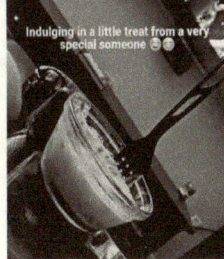

There's something peaceful about this place 😊

About the Author

Hugo Bradford is originally from El Paso, TX, where he attended college at UTEP, and his family still resides. He's lived in Houston, TX, since 2012, a city he loves dearly for challenging him, for its green and its amazing people.

Hugo spends most of his days learning about life's many wonders and scientific beauties while working from home on his two businesses, Voguish Apparel (www.VoguishApparel.com,) and (www.CanvasMafia.com), dedicating his life to art. One, a nationally trademarked women's clothing line and Canvas Mafia, an entrepreneurial canvas art business, motivating the businessmen and women of tomorrow one person at a time. He also dedicates his time to writing since his author's debut and looks forward to his future projects.

Upcoming Projects

Un Invierno Azul Violeta

This fictional love story is written in Hugo's native language as a "sonnet" for his Spanish speaking followers, recounting the tale of Gael and Alondra's impossible love. A story that pays homage to the author's beginnings on the subject of love, following "The Anatomy of Love" with a poetic, enthralling tale.

Unplugged

Following on his accolades as a science writer, Hugo looks to unplug you from the notion of the "status quo." A book with grit that touches on controversial subjects such as the banking system, racial issues, big pharma, and many more.

*F*** Your Feelings*

A non-fiction book that debuts the author's untamed voice, as it dives deep into the abyss of a generation's change of tone. A book not meant for the easily offended, as Hugo speaks like he never has before in an attempt to challenge the voice of reason of the present, uncaring of the appeal to the masses.

Finding the Author

On Instagram: www.instagram.com/iamhugobradford

Facebook: www.facebook.com/authorhugobradford

YouTube:

https://www.youtube.com/channel/UCgwoJRD4rmRjf3SfM5E73PQ?

Or on his website: www.HugoBradford.com

Feel free to write Hugo at Hugo@theanatomyof.love or with an eye-catching direct message on his Instagram page, as he would love to hear from you.

I want to thank every single one of you from the bottom of my heart, as your love and support has been incredibly humbling, and if not for you, this page and my life would not be possible. Thank you for reading my book, and thank you again and again for your inspiring affection. I look forward to entertaining you and, hopefully, teaching you new things for many years to come.

www.ingramcontent.com/pod-product-compliance
Lightning Source LLC
Chambersburg PA
CBHW030637150426
42811CB00083B/2395/J